In Defence of Objective Bayesianism

Jon Williamson
University of Kent

Great Clarendon Street, Oxford OX2 6DP

Oxford University Press is a department of the University of Oxford.
It furthers the University's objective of excellence in research, scholarship,
and education by publishing worldwide in

Oxford New York

Auckland Cape Town Dar es Salaam Hong Kong Karachi
Kuala Lumpur Madrid Melbourne Mexico City Nairobi
New Delhi Shanghai Taipei Toronto

With offices in

Argentina Austria Brazil Chile Czech Republic France Greece
Guatemala Hungary Italy Japan Poland Portugal Singapore
South Korea Switzerland Thailand Turkey Ukraine Vietnam

Oxford is a registered trade mark of Oxford University Press
in the UK and in certain other countries

Published in the United States
by Oxford University Press Inc., New York

© Jon Williamson 2010

The moral rights of the author have been asserted
Database right Oxford University Press (maker)

First published 2010

All rights reserved. No part of this publication may be reproduced,
stored in a retrieval system, or transmitted, in any form or by any means,
without the prior permission in writing of Oxford University Press,
or as expressly permitted by law, or under terms agreed with the appropriate
reprographics rights organization. Enquiries concerning reproduction
outside the scope of the above should be sent to the Rights Department,
Oxford University Press, at the address above

You must not circulate this book in any other binding or cover
and you must impose the same condition on any acquirer

British Library Cataloguing in Publication Data

Data available

Library of Congress Cataloging in Publication Data

Data available

Typeset by SPI Publisher Services, Pondicherry, India
Printed in Great Britain
on acid-free paper by
the MPG Books Group, Bodmin and King's Lynn

ISBN 978-0-19-922800-3

1 3 5 7 9 10 8 6 4 2

Preface

Bayesian epistemology aims to answer the following question: How strongly should an agent believe the various propositions expressible in her language? Subjective Bayesians hold that it is largely (though not entirely) up to the agent as to which degrees of belief to adopt. Objective Bayesians, on the other hand, maintain that appropriate degrees of belief are largely (though not entirely) determined by the agent's evidence. This book states and defends a version of objective Bayesian epistemology. According to this version, objective Bayesianism is characterized by three norms:

Probability: degrees of belief should be probabilities,

Calibration: they should be calibrated with evidence,

Equivocation: they should otherwise equivocate between basic outcomes.

The book is structured as follows. Chapter 1 introduces the main ideas behind objective Bayesianism and sets the scope of the book. Chapter 2 places objective Bayesian epistemology within the context of the quest for a viable interpretation of probability; it then provides the core characterization of objective Bayesianism that will be developed and defended in the remainder of the book. Chapter 3 presents some of the motivation behind the norms of objective Bayesianism. Chapter 4 discusses the differences between methods for updating objective and subjective Bayesian degrees of belief. Chapter 5 extends the characterization of objective Bayesianism from propositional to predicate languages. Chapter 6 develops computational tools for objective Bayesianism and considers an application of the resulting methods to cancer prognosis. Chapter 7 applies objective Bayesianism to the task of providing a semantics for probabilistic logic. Chapter 8 applies objective Bayesianism to the problem of judgement aggregation. Chapter 9 extends objective Bayesianism to richer languages—in particular the language of the mathematical theory of probability—and considers concerns about language relativity and objectivity. Chapter 10 develops some questions and topics for further research.

Familiarity with logical and mathematical notation is assumed, but no advanced knowledge of either logic or probability theory.

The reader intending merely to dip into the book is advised to read §2.3 first, which sets out some of the principal ideas and notational conventions that are applied throughout the book. For ease of reference, the key notation is summarized as follows. An agent's language is denoted by $\mathcal{L}$ and her evidence by $\mathcal{E}$; her evidence includes everything that she takes for granted in her current operating context. The agent's belief function over $\mathcal{L}$ is denoted by $P_{\mathcal{E}}^{\mathcal{L}}$ or simply $P_{\mathcal{E}}$; $P_{\mathcal{E}}(\theta)$ is the degree to which the agent believes sentence θ of $\mathcal{L}$. It is assumed that $\mathcal{E}$ imposes certain constraints χ that the agent's belief function ought to satisfy.

The set of probability functions satisfying these constraints is denoted by $\mathbb{E}$. In this book, $\mathbb{E}$ is taken to be $\langle \mathbb{P}^* \rangle \cap \mathbb{S}$, where $\langle \mathbb{P}^* \rangle$ is the convex hull of the set $\mathbb{P}^*$ within which evidence suggests that the chance function P^* lies, and $\mathbb{S}$ is a set of probability functions satisfying structural constraints imposed by qualitative evidence. Objective Bayesianism is taken to endorse the selection of a belief function $P_\mathcal{E} \in \Downarrow\mathbb{E}$, where $\Downarrow\mathbb{E}$ is the set of functions in $\mathbb{E}$ that are sufficiently close to a maximally equivocal function $P_=$, called the equivocator on $\mathcal{L}$. (If there is no single maximally equivocal function, then objective Bayesianism requires that $P_\mathcal{E}$ be sufficiently close to the set $\mathbb{P}_=$ of potential equivocators.) $\Downarrow\mathbb{E}$ is non-empty and $\downarrow\mathbb{E} \subseteq \Downarrow\mathbb{E} \subseteq \mathbb{E}$, where $\downarrow\mathbb{E}$ is the set of probability functions in $\mathbb{E}$ that are closest to the equivocator.

The research presented here pulls together ideas from a number of earlier works. Chapter 3 develops the approach of Williamson (2005a, 2007b, 2009b). The argument presented in Chapter 4 appeared in Williamson (2009b). Chapter 5 develops an approach taken in Williamson (2008a) and supersedes an alternative approach presented in Williamson (2008b, 2007a). The objective Bayesian net (obnet) formalism of Chapter 6 was developed in Williamson (2001, 2002, 2005a,b, 2008a); the application to cancer is explored further in Nagl et al. (2008) and Russo and Williamson (2007). Chapter 7 presents the core ideas of Haenni et al. (2010), to which the interested reader is urged to refer for more detail. Chapter 8 presents material in Williamson (2009a). Chapter 9 develops themes in Williamson (2009c).

I am extremely grateful for the help I have received in preparing this book. I have benefitted from too many discussions and exchanges to list here, but in particular Donald Gillies, Federica Russo and Gregory Wheeler provided very useful comments on an earlier draft. The book was made possible by a Leverhulme Trust Research Fellowship, by the ever-helpful mathematics editors at Oxford University Press—including Helen Eaton, Dewi Jackson, and Alison Jones—and by the patience and constant support of my family.

Contents

1 Introduction — 1
  1.1 Objective Bayesianism outlined — 1
  1.2 Objective Bayesian theory — 2
  1.3 Criticisms of objective Bayesianism — 3
  1.4 Evidence, language, and rationality — 4

2 Objective Bayesianism — 10
  2.1 Desiderata for a theory of probability — 10
  2.2 From Jakob Bernoulli to Edwin Jaynes — 12
  2.3 A characterization of objective Bayesianism — 26

3 Motivation — 31
  3.1 Beliefs and bets — 31
  3.2 Probability — 33
  3.3 Calibration — 39
  3.4 Equivocation — 49
  3.5 Radical subjectivism — 72

4 Updating — 75
  4.1 Objective and subjective Bayesian updating — 75
  4.2 Four kinds of incompatibility — 78
  4.3 Criticisms of conditionalization — 82
  4.4 A Dutch book for conditionalization? — 85
  4.5 Conditionalization from conservativity? — 88
  4.6 Summary — 89

5 Predicate Languages — 90
  5.1 The framework — 90
  5.2 The Probability norm — 92
  5.3 Properties of the closer relation — 95
  5.4 Closure — 96
  5.5 Characterizing Equivocation — 98
  5.6 Order invariance — 101
  5.7 Equidistance — 103

6 Objective Bayesian Nets — 108
  6.1 Probabilistic networks — 108
  6.2 Representing objective Bayesian probability — 112
  6.3 Application to cancer prognosis — 116

7 Probabilistic Logic — 121
  7.1 A formal framework for probabilistic logics — 121
  7.2 A range of semantics — 123

	7.3	Objective Bayesian semantics	129
	7.4	A calculus for the objective Bayesian semantics	133
8	**Judgement Aggregation**		136
	8.1	Aggregating judgements	136
	8.2	Belief revision and merging	137
	8.3	Merging evidence	139
	8.4	From merged evidence to judgements	142
	8.5	Discussion	144
9	**Languages and Relativity**		148
	9.1	Richer languages	148
	9.2	Language relativity	155
	9.3	Objectivity	157
10	**Objective Bayesianism in Perspective**		163
	10.1	The state of play	163
	10.2	Statistics	165
	10.3	Confirmation and science	169
	10.4	Epistemic metaphysics	170
	References		173
	Index		183

1
Introduction

We begin in §1.1 with an informal and intuitive introduction to objective Bayesianism as the view that an agent's degrees of belief should be probabilities, should respect constraints imposed by empirical evidence, but should otherwise equivocate between basic propositions. This is followed in §1.2 by an outline of the positive claims of the book and in §1.3 by an overview of the key challenges and objections faced by objective Bayesianism, and pointers to where in the book each of these objections are met. Then §1.4 sets out some of the key assumptions of the book concerning an agent's evidence, her language, and rationality.

1.1 Objective Bayesianism outlined

Objective Bayesian epistemology (also known simply as *objective Bayesianism*) is a theory about how strongly one should believe propositions that are open to speculation. It is a normative theory, concerning the strength with which one *ought* to believe certain propositions, rather than a descriptive theory about one's actual degrees of belief. This book presents a version of objective Bayesianism that is characterized by the following tenets. The strengths of an agent's beliefs should behave like probabilities: they should be representable by real numbers in the unit interval and one should believe a disjunction of mutually exclusive propositions to the extent of the sum of the degrees of belief of the disjuncts. Moreover, these degrees of belief should be shaped by empirical evidence: for example, they should be calibrated with known frequencies. Where empirical evidence does not fully determine degrees of belief, they should equivocate. Thus, the version of objective Bayesianism presented here can be summed up as follows: *degrees of belief should be probabilistic, calibrated with evidence, and otherwise equivocal.*

Consider, for example, a consultant who needs to determine a prognosis for a patient who has breast cancer. The consultant knows only that the patient has been treated for breast cancer, and that in somewhere between 10% and 40% of such cases the cancer recurs. Objective Bayesianism then maintains the following. The degree to which the agent should believe that the patient's cancer will recur (R) is representable by a probability $P(R) \in [0, 1]$, where $P(R) + P(\neg R) = 1$. This degree of belief should be constrained by the frequency evidence, that is, $P(R) \in [0.1, 0.4]$. Finally, the consultant should equivocate as far as possible between R and $\neg R$, and therefore should set $P(R) = 0.4$ and $P(\neg R) = 0.6$.

The aim of this book is to provide a detailed development of objective Bayesianism and to defend the resulting theory against a broad range of criticisms.

1.2 Objective Bayesian theory

This book is as much about theory-building as about meeting objections to objective Bayesianism. Here we set the scope of objective Bayesian theory as developed in the book.

The orthodox view of Bayesian epistemology goes something like this. All Bayesian epistemologists hold that rational degrees of belief are probabilities, consistent with total available evidence, and updated in the light of new evidence by Bayesian conditionalization. Strict subjectivists (e.g. Bruno de Finetti) hold that initial or *prior* degrees of belief are largely a question of personal choice. Empirically based subjectivists (e.g. Howson and Urbach 1989*a*) hold that prior degrees of belief should not only be consistent with total evidence, but should also be calibrated with physical probabilities to the extent that they are known. Objectivists (e.g. Edwin Jaynes) hold that prior degrees of belief are fully determined by the evidence.

This book preserves the distinction between the three kinds of Bayesian epistemologies, but departs from orthodoxy inasmuch as it develops a version of objective Bayesianism that is not updated by conditionalization, and where probabilities are not fully determined by evidence—they are not always fully determined, and where they are, they are determined by more than evidence alone.

According to the version of objective Bayesian epistemology developed here, an agent's degrees of belief should depend on her evidence, her language, and various other contextual factors. As explained in §1.4, the agent's evidence includes everything she takes for granted in her current operating context. Her language not only allows her to express the propositions of interest to her, but also conceptualizes the salient features of her world and delimits the range of possible circumstances that she can consider. We restrict our attention to three kinds of formal languages in this book: the language of propositional logic (§2.3), the language of predicate logic (§5.1), and the language of the mathematical theory of probability (§9.1).

In this book, objective Bayesianism is characterized in terms of three norms. The Probability norm says that the strengths of an agent's beliefs should be representable by a probability function defined over the sentences of the agent's language. The Calibration norm says that this probability function should fit with the agent's evidence. The Equivocation norm says that, to the extent that evidence leaves open a range of evidentially compatible probability functions, the agent's degrees of belief should be representable by some such function that equivocates to a sufficient extent between the basic propositions expressible in her language. The different versions of Bayesianism are distinguishable according to their stance regarding these norms: while objective Bayesianism advocates all three norms, empirically based subjectivists accept Probability and Calibration but reject Equivocation, and strict subjectivists only endorse the Probability norm. The book explicates these norms in a more precise way and shows that they do indeed yield strong constraints on degrees of belief. Chapter 3 discusses the motivation behind the three norms, arguing that any attempt at rigorously proving a norm will be futile in the sense that such a proof is unlikely to convince a detractor. However, it is argued that the norms do admit compelling pragmatic justifications. So, while the norms cannot be proven and need to be suitably qualified, all three norms are genuine constraints on rational belief and not merely conventions (§3.5).

The book focuses on explicating the Equivocation norm through the versions of Edwin Jaynes' Maximum Entropy Principle (see §2.3), although §3.4 leaves open the possibility of other explications. This yields a natural mechanism for updating degrees of belief which differs from the subjective Bayesian method of updating, Bayesian conditionalization (Chapter 4). It also makes possible a relatively efficient way of calculating objective Bayesian probabilities, the machinery of objective Bayesian nets developed in Chapter 6.

Although in its infancy in comparison with other interpretations of probability (§2.2), objective Bayesian theory already admits many applications, for example to decision support systems (§6.3), to logic (Chapter 7), to epistemology (Chapter 8), to statistics (§10.2), to scientific reasoning (§10.3), and to metaphysics (§10.4). It is hoped that the book will encourage others to further develop objective Bayesian theory and its applications.

1.3 Criticisms of objective Bayesianism

Objective Bayesianism has been challenged on a number of different fronts.

It has been accused of being poorly motivated. For example, Festa (1993, §7.2) criticizes the Maximum Entropy Principle on the grounds that it is not clear how evidence constrains probability. It is hoped that by being explicit about the three norms of objective Bayesianism, and being explicit about the extent and limitations of the motivation behind each of these norms (Chapter 3), many of these worries can be allayed.

Objective Bayesianism has been accused of poorly handling qualitative evidence. For example, Pearl (1988, p. 463) maintains that the Maximum Entropy Principle fails to adequately handle causal evidence. This criticism is addressed in §3.3 as well as in Williamson (2005a). No doubt there is more to do in this direction, but as things stand various kinds of qualitative evidence—including evidence of causal, logical, semantic, and ontological influence—can now be handled.

Objective Bayesianism has been accused of yielding updates that can disagree with Bayesian conditionalization (see, e.g. Friedman and Shimony 1971). It is shown in Chapter 4 that while this is indeed the case, it reflects negatively on Bayesian conditionalization rather than on objective Bayesian updating.

Objective Bayesianism has been accused of suffering from a failure to learn from experience (see, e.g. Dias and Shimony 1981, §4; Paris 1994, p. 178). It is shown in §4.3 that this objection rests on a mistaken understanding of objective Bayesian conditional probabilities.

Objective Bayesianism has been accused of being computationally intractable. For instance, Pearl (1988, p. 463) dismisses the Maximum Entropy

Principle on these grounds. Chapter 6 shows that Pearl's own Bayesian network formalism can achieve a significant dimension reduction in the entropy maximization task, and can render objective Bayesianism computationally tractable.

Objective Bayesianism has been accused of being susceptible to the paradoxes of the Principle of Indifference: there may be more than one partition of sentences over which one might equivocate, each partition giving different conclusions (see, e.g. Seidenfeld 1986, §3). In §9.1, it is acknowledged that there may in certain cases be more than one potential equivocator—a function that is maximally equivocal over the basic propositions expressible in the language—but at worst this leads to subjectivity rather than paradox.

Relatedly, objective Bayesianism has been accused of being language-dependent, that is, of yielding probabilities that depend on an agent's language as well as her evidence (see, e.g. Seidenfeld 1986, §3). This is indeed the case, but only to be expected, since an agent's language says something about the world in which she lives—it embodies implicit evidence (§9.2).

One might object to objective Bayesianism that it is not objective enough: there are several situations in which the three norms fail to determine a unique probability function, and, in any case, objective Bayesian probability is relative to evidence and language. In fact, §9.3 shows that subjectivity in the former sense is by no means pernicious and that relativity to evidence and language can be eliminated if need be.

No doubt there are some criticisms of objective Bayesianism in the literature that do not fall under one of the aforementioned broad objections. But those set out above have been the most influential criticisms and are the principal reasons invoked by subjectivists for preferring subjective over objective Bayesian epistemology. By showing that these objections can be substantially met, the book aims to level the playing field.

The book contains no sustained attack on subjectivism, many of which have been conducted elsewhere, but in defending objective Bayesianism several considerations that cast doubt on subjectivism will emerge.

1.4 Evidence, language, and rationality

In this section, we set out and discuss some of the presuppositions that pervade the rest of the book.

1.4.1 *Evidence*

Objective Bayesianism posits a link between an agent's evidence and her rational degrees of belief: degrees of belief should be constrained by the evidence and otherwise maximally equivocal. *Evidence* is understood here in a broad sense to include *everything the agent takes for granted* in her current operating context—for example, observations, theory, background knowledge, and assumptions. The terms 'total evidence' or 'epistemic background' or 'data' are sometimes used to refer to what is called 'evidence' in this book.

Note that an agent's evidence is relative to her current operating context. A medical consultant, for example, is rational to take for granted much of her education and medical training as well as uncontroversial studies in the medical literature and observations of the patient's symptoms, given her purpose of treating a patient. On the other hand, the purposes of a philosophical sceptic preclude her from taking much for granted at all. Therefore, if the medical consultant by day studies philosophy by night, her evidence base may change radically.

The agent's evidence is denoted by the symbol $\mathcal{E}$, and the probability function that represents the degrees of belief that the agent should adopt is denoted by $P_{\mathcal{E}}$. There is no need to assume that the evidence $\mathcal{E}$ is a set of propositions, or that the agent is always aware of what she grants—we can leave these questions open. Indeed, since objective Bayesianism only concerns the *link between* $\mathcal{E}$ and $P_{\mathcal{E}}$, there is no need to say very much about $\mathcal{E}$ at all, other than to explain how it constrains $P_{\mathcal{E}}$. Note, in particular, that an agent may or may not be rational to take $\mathcal{E}$ for granted, but objective Bayesianism says nothing about this question. Indeed, it is important to emphasize that the theory of objective Bayesianism does not aim to answer all questions about epistemic rationality: it is but one piece in the jigsaw puzzle of normativity.

It is also important to emphasize that an agent's evidence includes more than her background knowledge. Evidence (understood as everything the agent takes for granted in her current context) differs from knowledge in three respects. First, an agent may take something for granted when she ought not: for example, she may take some testimony for granted although she is aware that it is probably idle gossip. Such evidence does not qualify as knowledge because it lacks the required justification, or because it is arrived at through an unreliable or otherwise inappropriate process.[1] Second, an agent may take something for granted—and be justified in so doing—yet this something fails to qualify as knowledge because it is false: for example, she may take some testimony from a very reliable witness for granted, but it just turns out to be a rare occasion in which the witness was mistaken. Third, it may be the case that the agent does not take an item of her knowledge for granted: perhaps she should have taken it for granted but didn't, or perhaps she was right not to take it for granted because it was irrelevant in her operating context or because her operating context placed more stringent demands than does knowledge (e.g. that a reliable witness makes a true claim may be sufficient for that claim to be counted as knowledge, but not sufficient for it to be counted as evidence presentable to a court of law).

Not everyone agrees that evidence (taken as that which grounds belief) and knowledge differ. For example, Timothy Williamson maintains that knowledge and evidence are coextensive (Williamson 2000, chapter 9). Williamson

---

[1] Of course it is controversial as to how knowledge itself should be understood. However, accounts of knowledge standardly maintain that truth is necessary but insufficient for knowledge, so that there are at least two ways in which a knowledge claim might be false—it may lack truth or may lack whatever else is required for it to count as knowledge. It is this 'whatever else' that is intended here.

suggests two arguments for the claim that evidence must be true. First, 'if one's evidence included falsehoods, it would rule out some truths, by being inconsistent with them. One's evidence may make some truths improbable, but it should not exclude any outright' (Williamson 2000, p. 201). In response, one might agree that it is desirable that all one's evidence be true, but point out that mere desirability does not make it so. A possible second argument concerns a different *relational* notion of evidence—evidence *for* a hypothesis:

1: 'If e is evidence for h, then e is true. .... For example, that the ground is wet is evidence that it rained last night only if the ground is wet' (Williamson 2000, p. 201).

A link is posited between the two notions of evidence:

2: e is evidence for h for an agent if and only if it is in the agent's body of evidence, and e raises the probability of h (Williamson 2000, p. 187).

Suppose that one can make a case that

3: if e is in the agent's body of evidence, then there is some h whose probability e raises.

Then one could conclude that e being in the agent's body of evidence implies that it is true. In response to this second argument one can challenge any one of its premises. Consider the claim, common in discussions of computational complexity in computer science, that $P \neq NP$. This statement has not been proven and it is universally agreed that it might be false, yet the overwhelming majority of computer scientists grant it, and grant it for many good reasons. Now consider premise 1. It is commonly agreed that $P \neq NP$ is evidence that the travelling salesman problem is computationally intractable. Now $P \neq NP$ may be false. Should it be found that $P = NP$, then arguably the evidential claim—the claim that $P \neq NP$ is evidence for the travelling salesman problem being intractable—would be considered irrelevant rather than false. It would be what Williamson calls 'conversationally inappropriate' to utter such a claim, and in response to such a claim one would point to $P \neq NP$ being false, rather than the claim as a whole being false. To rebut premise 2, note that $P \neq NP$ does not raise the probability of the travelling salesman problem being intractable, on any standard account of probability including Williamson's own and that of this book, since any standard account gives $P \neq NP$ probability 0 or 1. Indeed, $P \neq NP$ raises the probability of no proposition so premise 3 must be false. In sum, neither of these two arguments forces the conclusion that evidence is knowledge.

Discussions of Bayesian epistemology in the literature often take background knowledge to be the sole constrainer of degrees of belief. In fact, though, one's degrees of belief are—and should be—guided by what one grants rather than what one knows. If knowledge ought to guide an agent when she is consciously deciding how strongly to believe a proposition that is open to speculation, then, since *ought* implies *can*, the agent should be able to identify exactly what it is she knows. But of course an agent cannot be expected to do this, because while she may think she knows something, she may be mistaken about its justification or its truth. On the other hand, the agent can identify what she takes for granted (at least with some training and time to reflect) and so can use her evidence to guide the strengths of her beliefs. Of course *can* does not imply *ought*: that evidence (taken as whatever the agent grants) can guide belief does not imply that it should. In particular, while one may accept that knowledge is too strong a condition on the grounds for belief, one might also suspect that evidence is too strong a condition. Might one not base belief on something weaker than what is *taken for granted*? On other beliefs perhaps? Thus, believing a proposition strongly (without taking it for granted) may be grounds for believing its negation weakly. In response to this worry it suffices to point out that while the proposition itself is not granted and hence does not qualify as evidence, *believing the proposition strongly* is granted and so does qualify as evidence. Hence, the *grounds* for believing the negation of the proposition weakly are granted rather than merely believed. In general, the grounds for one's beliefs need to be taken for granted if they are to ground those beliefs at all: if X warrants epistemic state E and one adopts state E on the basis of X, then one had better grant X. Consequently, it is evidence (construed as that which one grants), not something stronger or weaker, that should guide the strengths of one's beliefs.

That it is evidence not knowledge that should ground belief is apparent from examples in the legal setting where much effort has been directed at understanding belief formation. In a court of law the jurors should apportion their beliefs not according to their knowledge but according to the evidence under consideration: some of this evidence may in fact be false and hence not count as knowledge; for it to guide the jurors' beliefs, it is sufficient that they grant it. On the other hand, for technical reasons the judge may instruct that something the jurors have heard and known be discounted as evidence and should not influence their beliefs. It is not just belief formation in the law that behaves like this: in most contexts it is appropriate that our observations guide our beliefs, even those that turn out to be erroneous; in many contexts in engineering and science it is appropriate to grant Newtonian mechanics, even though it is known to be strictly false; an observation of a planet by the naked eye may yield knowledge of the planet, yet in the context of astronomical theorizing more precise and sure observation may be required in order for that knowledge to be granted, and hence guide the ascription of belief.

1.4.2 *Language*

While we can leave open the possibility that elements of the agent's evidence be non-propositional, it will be assumed that the objects of the agent's degrees of belief are propositions. It is the agent's *language*, which we will usually denote by $\mathcal{L}$, that picks out these propositions. Just as evidence may vary according to the agent's operating context, so too her language may vary with context. Thus, a structural engineer who moonlights as a baker may be thought of as having two languages whose technical terms intersect little. We need not assume anything about the relation between degrees of beliefs on different languages of the same agent where the context varies; if the languages intersect, one may want degrees of belief on the different languages to be consistent on their intersection, or some weaker paraconsistency may be sufficient, but there is no need to decide such issues here. On the other hand, the methods of this book will apply to the case in which an agent's language *in a single context* changes over time.

It will be helpful for the most part to consider formal languages: we shall assume that if the language can express propositions θ, φ then it can express $\neg \theta$ (not θ), $\theta \wedge \varphi$ (θ and φ), $\theta \vee \varphi$ (θ or φ), $\theta \rightarrow \varphi$ (not θ, or φ), and $\theta \leftrightarrow \varphi$ (both or neither of θ and φ). We shall at various points of this book consider languages that have finitely many, or denumerably many, or uncountably many elementary (i.e. non-logically complex) sentences, and we shall in some cases exploit the fine-grained structure of sentences by considering first-order logical languages with predicate, constant, variable, and quantifier symbols.

To say that an agent has language $\mathcal{L}$ and evidence $\mathcal{E}$ is not to say that there is no relation between evidence and language. While we do not presume that an agent's evidence $\mathcal{E}$ can be expressed in her language $\mathcal{L}$, we do not rule this out. Moreover, in Chapter 9 we shall see that the agent's language itself constitutes a kind of implicit evidence, conveying information about the world in which she operates. Thus, in principle the agent's language as well as her other evidence $\mathcal{E}$ can constrain her degrees of belief. Where we need to emphasize this dependence of degrees of belief on language as well as evidence, we shall denote the agent's belief function by $P_{\mathcal{E}}^{\mathcal{L}}$.

1.4.3 *Rationality*

This book focuses on *rational* degree of belief. Here 'rational' may simply be taken to mean *fit-for-purpose*. (These terms are clearly not inter-substitutable, though, because they have different domains of application: 'rational' is a normative notion and is applied to things over which we have some control—such as the strengths of our beliefs in propositions that are open to speculation—while 'fit-for-purpose' is descriptive and more widely applicable, for example, to artifacts, such as tables and chairs.)

One's degrees of belief are rational inasmuch as they are fit for the uses to which they are put. These uses are principally truth-orientated—for example, making predictions that turn out true, assessing which diagnosis is the right one, and determining the correct explanation. This book will only be concerned with degrees of belief in the context of these truth-orientated uses—these uses broadly require that degrees of belief fit reality as far as possible. Particular individuals may have ulterior motives when deciding what to believe. For instance, if one wants beliefs to lead to happiness, self-delusion may be the appropriate strategy in so far as it is possible to carry out. Interesting as these ancillary uses are, they will not be pursued here.

The type of rationality under consideration here is sometimes called *theoretical rationality* or *epistemic rationality* to distinguish it from *practical rationality* or *pragmatic rationality*. There are some grounds for drawing a distinction here (Harman 2004), but it must be drawn with caution. In particular, Gilbert Harman goes perhaps too far by arguing that theoretical reasoning is not characterizable as practical reasoning with a theoretical goal, on the grounds that practical reasoning involves arbitrary decisions (in deciding how to achieve a result there may be several equally promising means to that end, requiring an arbitrary choice of which means to adopt), while theoretical reasoning does not (Harman 2007, §4). We shall see in this book that theoretical reasoning *can* involve arbitrary choice: there may be several equally rational degrees of belief from which the agent should choose one arbitrarily. Furthermore, we shall see that even though the *uses* of degrees of belief may be truth-orientated and hence there is a sense in which reasoning with rational degrees of belief qualifies as theoretical reasoning, this kind of reasoning can also be characterized as reasoning with a *practical goal* and hence there is also a sense in which it is practical reasoning. This is because for degrees of belief to be deemed rational, they need to avoid the possibility of certain loss in both the short term (§3.2) and the long term (§3.3), and to minimize worst-case expected loss (§3.4), all of which are practical ends. In general, while many attempt to drive a wedge between belief and action (see, e.g. Kyburg Jr 1970), they are but two sides of the same coin—beliefs are the epistemic determinants of action—and one can straightforwardly draw inferences from one to the other (§§3.1, 3.5). This truism helps to muddy the water in distinguishing theoretical from practical rationality, a phenomenon known as *pragmatic encroachment* (see, e.g. Fantl and McGrath 2007). Pragmatic encroachment need not detain us here. By embracing the pragmatic aspects of belief but distinguishing its truth-orientated *uses* from ulterior motives, we have enough of a distinction to focus on truth-orientated rationality in this book.

In sum, rational degree of belief is relative to evidence and language. Degrees of belief are rational if they are fit-for-purpose. Plausibly, degrees of belief are rational if and only if they are determined in the right way from evidence and language, and if in the first place the agent is rational to grant that evidence and adopt that language (in so far as the agent has any choice about her evidence and language). Objective Bayesian epistemology concerns only the former question: the link between evidence and language on the one hand, and rational degree of belief on the other; it does not concern choice of evidence or choice of language. Objective Bayesianism is a theory which holds that the degrees of belief which best fit their purpose are those that are probabilistic, calibrated with evidence, and otherwise equivocal. In Chapter 3, we examine some of the reasons for invoking these principles, but first we shall take a look at some of the key developments on the path to present-day objective Bayesianism.

2
Objective Bayesianism

This chapter reviews the various interpretations of probability (§2.2), assessing them in the light of certain desiderata for a philosophical theory of probability that are presented in §2.1. It is suggested that present-day objective Bayesianism, which is characterized in §2.3, is close to early views of probability, such as those of Jakob Bernoulli and Thomas Bayes.

2.1 Desiderata for a theory of probability

This book is primarily concerned with objective Bayesian *epistemology*: objective Bayesianism construed as a theory about rational belief, which holds that one's degrees of belief should be certain probabilities. But much of the literature on objective Bayesianism is concerned with the converse question: should probabilities be interpreted as rational degrees of belief? Clearly, the two questions are related. If probabilities should be interpreted as rational degrees of belief then these rational degrees of belief are probabilities. On the other hand, if rational degrees of belief are probabilities, then at least those probabilities are interpretable as rational degrees of belief. But in principle one can advocate objective Bayesian epistemology without maintaining that probabilities should always, or even often, be interpreted as rational degrees of belief. The *objective Bayesian interpretation of probability* is the construal of probabilities in terms of rational degrees of belief, where these rational degrees of belief are given an objective Bayesian account.

In order to understand the historical development of objective Bayesianism, we will need to understand the quest for a viable interpretation of probability. For an interpretation or philosophical theory of probability to be considered viable, there are a number of requirements that need to be fulfilled. While opinions differ as to exactly what these requirements might be, the following might be considered core requirements. A philosophical interpretation of probability should

Objectivity: account for the objectivity of probability,

Calculi: explain how we reason about probability,

Epistemology: explain how we can know about probability,

Variety: cope with the full variety of probabilistic claims that we make, and

Parsimony: be ontologically parsimonious.

It should be stressed that these are requirements made of a philosophical theory of probability in order for that theory to be judged a viable *interpretation* of probabilistic statements as they are routinely used. They are not requirements made of a philosophical theory of probability in its role as providing an account of degrees of belief, for example. In the latter context, conditions such as Objectivity and Parsimony are far from compelling. Note that Salmon (1967, §IV.2) put forward three desiderata for an interpretation of probability: Admissibility, Ascertainability, and Applicability, which correspond roughly to our desiderata of Calculi, Epistemology, and Variety, respectively.

Let us consider these desiderata in turn.

2.1.1 Objectivity

Many applications of probability invoke a notion of probability that is objective in a logical sense: there is a fact of the matter as to what the probabilities are; if two agents disagree about a probability, at least one of them must be wrong. (Logical objectivity contrasts with the ontological sense of objectivity: probabilities are ontologically objective if they exist as entities or are reducible to existing entities, and are ontologically independent of mental or epistemological considerations.) For example, the probability that a patient's breast cancer will recur after treatment apparently depends on features of the cancer, of the treatment, and of the patient. It is not simply a matter of personal opinion: if two prognostic probabilities differ, at least one of them must be wrong. A philosophical interpretation of probability should, if possible, yield a notion of probability that is suitably objective in this logical sense—otherwise, it is revising rather than faithfully interpreting probabilistic statements as they occur in these applications.

2.1.2 Calculi

Probabilities are manipulated and inferences are drawn from them by means of the probability calculus. This mathematical apparatus, based on axioms put forward by Kolmogorov (1933) has by now become well entrenched. Consequently, a philosophical interpretation of probability should yield a notion that satisfies the axioms of probability. Otherwise, it is not a theory of probability—it is a theory of something else.

The mathematical probability calculus proceeds along the following lines. Given an *outcome space* Ω which represents the possible outcomes of an experiment or observation, an *event space* is a set of subsets of Ω. An event—a subset of Ω—represents the proposition that the actual outcome lies in that set of outcomes. An event space $\mathcal{F}$ is a *field* if it contains Ω and is closed under the formation of complements and finite unions. A function P from a field $\mathcal{F}$ to the non-negative real numbers is a *probability function* if it satisfies Countable Additivity.

Countable Additivity: if $E_1, E_2, \ldots \in \mathcal{F}$ partition Ω (i.e. $E_i \cap E_j = \emptyset$ for $i \neq j$ and $\bigcup_{i=1}^{\infty} E_i = \Omega$), then $\sum_{i=1}^{\infty} P(E_i) = 1$.

In particular, $P(\Omega) = 1$ (take $\Omega, \emptyset, \emptyset, \emptyset, \ldots$ as the partition: $P(\emptyset) = 0$, for otherwise $\sum_{i=1}^{\infty} P(E_i) = \infty$; therefore $P(\Omega) = 1$). The triple $(\Omega, \mathcal{F}, P)$ is called a *probability space*.

2.1.3 *Epistemology*

We come to know about probabilities in various ways: we measure population frequencies, we appeal to symmetry arguments or scientific theories, we make educated guesses, and we derive some probabilities from others using the probability calculus. A philosophical theory of probability should explain how we can use such techniques to discover probabilities. If the theory rejects some of these techniques, it should say where they go wrong and why they are apparently successful.

2.1.4 *Variety*

Probabilistic claims are extremely varied. For instance, claims are made about single-case probabilities (e.g. the probability that a particular patient's cancer will recur) and generic probabilities (e.g. the probability of recurrence among those who receive radiotherapy). Moreover, probabilities are attached to a variety of entities, including events, sets, variables, sentences, propositions, and hypotheses. A philosophical theory of probability should be able to cope with this variety—it should account for each use of probability, or, if some uses are to be viewed as illegitimate, it should say how such uses should be eliminated in favour of the legitimate uses. Otherwise, the theory is at best a partial theory, a theory of some of the uses of probability.

2.1.5 *Parsimony*

Arguably a philosophical theory of probability should not make unwarranted ontological commitments: if one can reduce probabilities to something else in one's ontology, then one should do that rather than take probabilities as primitive. This is just Ockham's razor; it may be viewed as a methodological or psychological requirement and as such subsidiary to the other desiderata.

2.2 From Jakob Bernoulli to Edwin Jaynes

Having discussed what might be required of an interpretation of probability, we now turn to some of the key interpretations.

2.2.1 *Bernoulli's interpretation*

Jakob Bernoulli was perhaps the first to advocate the main tenets of what is now known as the objective Bayesian interpretation of probability. Bernoulli developed his view of probability in the 1680s, though it was not published until 1713, 8 years after his death. Bernoulli identified probability with degree of belief: '*Probability*, indeed, is degree of certainty, and differs from the latter as a part differs from the whole' (Bernoulli 1713, p. 211). Moreover, Bernoulli argued that these degrees of belief are determined by evidence, or, in the absence of evidence, by equivocating between possibilities:

> For example, three ships set sail from port. After some time it is reported that one of them has perished by shipwreck. Which do we conjecture it to be? If I considered only the number of ships, I might conclude that misfortune might equally befall any of them. But since I remember that one of them was more eaten away by decay and

age than the others, that it was badly equipped with sails and sail-yards, and also that it was commanded by a new and inexperienced skipper, I judge that it is surely more probable that this one perished than the others. (Bernoulli 1713, p. 215)

Let us see how Bernoulli's interpretation fares with respect to the desiderata outlined above.

It is not clear to what extent Bernoulli's concept of probability can be classed as objective (in the logical sense discussed in §2.1). On the one hand, probability is interpreted as degree of belief or certainty, and it seems plausible that different agents with the same evidence might believe the same event to different degrees. On the other hand, quotes like that given above suggest that probabilities are dependent on the extent and limitations of one's evidence, but not on other factors. Further argument is clearly required before one can be convinced that Bernoulli's interpretation satisfies the Objectivity desideratum.

Jakob Bernoulli was one of the pioneers of the mathematical notion of probability, proving a version of what we now call the weak law of large numbers. However, his probabilities are not probabilities in the modern sense since they were not additive: 'Thus it can happen that one thing has $\frac{2}{3}$ of certainty, while its contrary has $\frac{3}{4}$' (Bernoulli 1713, p. 221). Hence, Bernoulli's notion of probability does not satisfy the Calculi desideratum.

Bernoulli had a lot to say about the epistemology of probability. He insisted that probabilities should be based on all available evidence (Bernoulli 1713, pp. 214–15), and, as mentioned above, that probabilities are influenced by both the presence and absence of evidence. In the presence of evidence, Bernoulli advocated a principle akin to what is now known as the *Principle of the Narrowest Reference Class*:

> *Remote and universal arguments are sufficient for making judgements about universals, but when we make conjectures about individuals, we also need, if they are at all available, arguments that are closer and more particular to those individuals.* Thus when we are asked in the abstract how much more probable it is that a young man of twenty will outlive an old man of sixty rather than vice versa, there is nothing we can consider besides the difference in age. But when we discuss specific individuals, say the young man Peter and the old man Paul, we need to pay attention also to their particular constitutions and to the care that each one takes of his health. For if Peter were sickly, if he indulged his passions, or if he lived intemperately, then Paul, even though he is more advanced in age, might, with excellent reason, hope to live longer. (Bernoulli 1713, pp. 215–16)

Where the available evidence is inconclusive, Bernoulli maintained that one should only commit to the extent warranted by that evidence: 'In our judgements we should be careful not to attribute more weight to things than they have' (Bernoulli 1713, p. 216). Thus, Bernoulli provided the core of an epistemology of probability. This core is somewhat lacking in detail, and, as we shall see, lacked substantial further elaboration until the second half of the twentieth century.

Next, we turn to the desideratum of Variety. It should be noted that only probabilities of *single cases* can be directly interpreted as degrees of belief. Beliefs attach to single cases—propositions—not to generic outcomes such as open formulae or repeatable experiments. Thus, one can believe that the ship *Goliath* is wrecked, but one cannot believe that an *arbitrary* ship is wrecked, because nothing is being predicated of any particular ship. However, probabilities attach both to single cases and to repeatable or generic phenomena: one can sensibly say 'the probability that an arbitrary ship is wrecked in the coming year is 0.1'. There are at least three possible approaches to such generic statements that are open to the proponent of a rational-degree-of-belief (or *Bayesian*) interpretation of probability. She might say that generic probability statements are nonsense and should be abandoned. In that case, the Bayesian needs to explain why generic probability statements are used so widely, apparently sensibly, and apparently successfully. Alternatively, the Bayesian might say that generic probability statements talk not about the strength of a belief but about something else—for example, in this case perhaps about the proportion of ships that are wrecked. Then, the Bayesian would need to accept that a Bayesian interpretation can only directly interpret *some* probability statements, namely those which ascribe probabilities to single cases, and is not appropriate when dealing with generic probability statements. A third option is suggested by the above quotation of Bernoulli's. One might treat certain generic statements as referring to a single case where the evidence is necessarily limited. Thus, the probability that an *arbitrary* young man of 20 will outlive an *arbitrary* old man of 60 might be identified with the probability that Peter will outlive Paul, where all that is and can be known about Peter and Paul is that the former is 20 years old and the latter is 60. Then, generic probability statements can be indirectly interpreted as talking about rational degrees of belief. It is hard to say which of these three strategies would have been favoured by Bernoulli, but this last strategy seems a plausible reading, in which case Bernoulli's interpretation would cope well with the Variety desideratum.

Bayesian interpretations of probability fare well when it comes to Parsimony. Presumably, *any* metaphysics must make room for beliefs that vary in strength and indeed for rational degrees of belief. Hence, if probabilities are interpreted as rational degrees of belief, nothing need to be added to the ontology; there is no need to posit a new type of entity.

In sum, while Bernoulli's interpretation of probability has much in common with the objective Bayesian interpretation, it is not, as it stands, a viable interpretation. It is unclear as to how objective the interpretation is, and unclear as to the precise mechanisms for determining probabilities. It does not provide an interpretation of probability in its present, additive, sense, but, under a charitable reading of Bernoulli, it does interpret generic as well as single-case probability statements.[1]

---

[1] Hacking (1971) and Shafer (1996) also discuss Bernoulli's interpretation. Hacking suggests that Bernoulli's integration of frequency considerations and a degree-of-belief account of probability is rather incoherent: 'In fact Bernoulli was, like so many of us, attracted to many of these seemingly incompatible ideas, and he was unsure where to rest his case' (Hacking 1971, p. 210). Not so: contemporary objective Bayesianism, which is close in spirit to Bernoulli's own

2.2.2 Bayesian interpretations

Thomas Bayes' interpretation of probability retained the three core principles put forward by Jakob Bernoulli: (i) Probability is epistemic and normative, measuring rational expectation (Bayes 1764, Definition 1.5). (ii) Probability is partially determined by evidence, in particular evidence of observed frequencies (e.g. Bayes 1764, Rule 2). (iii) Probability is also partially determined by equivocating—not committing to one over any other number of successes in a sequence of trials (Bayes 1764, Scholium).

For reasons that are not entirely clear, what is now known as the *Bayesian interpretation of probability* encapsulates the first principle only. Indeed, any interpretation of probability that construes probabilities as rational degrees of belief is classed as 'Bayesian', although Bayes' view went well beyond this claim and in any case this claim is not original to Bayes. The Bayesian interpretation of probability should not be confused with *Bayesian epistemology* (which holds that the strengths of our beliefs should satisfy the axioms of probability), *Bayesian statistics* (which makes heavy use of Bayes' theorem, $P(\theta|\varphi) = P(\varphi|\theta)P(\theta)/P(\varphi)$—see §10.2), and *Bayesian confirmation theory* (which holds that a theory is confirmed by evidence to the extent of the probability of the theory given the evidence—see §10.3).[2]

Broadly speaking, there are three kinds of Bayesian epistemology and each kind gives rise to a distinct interpretation of probability. Any Bayesian epistemology holds that one's degrees of belief at a particular time must be probabilities if they are to be considered rational. We shall call this the *Probability* norm. Now *strictly subjective Bayesian epistemology* takes this condition to be sufficient for rationality as well as necessary: if your degrees of belief at a particular time are probabilities then they are rational. (This condition deals with rationality at a particular time; as we see in Chapter 4, subjective Bayesians also advocate *Bayesian conditionalization*, a rule for updating degrees of belief which holds that if one learns new evidence ε then one's new degrees of belief should match the old degrees conditional on E, $P'(\theta) = P(\theta|\varepsilon)$.) The strictly subjective Bayesian view was championed by de Finetti (1937); it leads naturally to a strictly subjective Bayesian interpretation of probability: probabilities are construed as rational degrees of belief where 'rational' is interpreted according to the norms of the strictly subjective Bayesian epistemology.

In contrast, *empirically based subjective Bayesian epistemology* takes the Probability norm together with a further *Calibration* norm as necessary and sufficient for rationality at a particular time: one's degrees of belief should not only be probabilities, they should also be calibrated with known frequencies. As Frank Ramsey writes

> Let us take a habit of forming opinion in a certain way; e.g. the habit of proceeding from the opinion that a toadstool is yellow to the opinion that it is unwholesome.

approach, shows that there is a coherent combination of frequency and degree of belief. It is clear that Bernoulli had latched on to the core ideas behind this combination.

[2] Fienberg (2006) offers an account of when Bayesian statistics became known as 'Bayesian'.

> Then we can accept the fact that the person has a habit of this sort, and merely ask what degree of opinion that the toadstool is unwholesome it would be best for him to entertain when he sees it; i.e. granting that he is doing to think always in the same way about all yellow toadstools, we can ask what degree of confidence it would be best for him to have that they are unwholesome. And the answer is that it will in general be best for his degree of belief that a yellow toadstool in unwholesome to be equal to the proportion of yellow toadstools which are in fact unwholesome. (Ramsey 1926, p. 50)

There is a corresponding empirically based subjective Bayesian interpretation of probability: probabilities are interpreted as rational degrees of belief, where 'rational' is cashed out via both the Calibration and Probability norms.

These two positions—strictly subjective Bayesianism and empirically based subjective Bayesianism—are both versions of what is known as *subjective Bayesian epistemology*. The third variety of Bayesian epistemology is *objective Bayesian epistemology*. This appeals to an *Equivocation* norm as well as the Calibration and Probability norms: one's degrees of belief at a particular time are rational if and only if they are probabilities, calibrated with physical probability and otherwise equivocate between the basic possibilities. Ramsey in fact advocated such a combination:

> When, however, the evidence is very complicated, statistics are introduced to simplify it. They are to be chosen in such a way as to influence me as nearly as possible in the same way as would the whole facts they represent if I could apprehend them clearly. But this cannot altogether be reduced to a formula; the rest of my knowledge may affect the matter; thus p may be equivalent in influence to q, but not ph to qh.
>
> There are exceptional cases in which 'It would be reasonable to think p' absolutely settles the matter. Thus if we are told that one of these people's names begins with A and that there are 8 of them, it is reasonable to believe to degree $\frac{1}{8}$-th that any particular one's name begins with A, and this is what we should all do (unless we felt that there was something else relevant). (Ramsey 1928, pp. 100–1)

This position is more in line with Bayes' and Bernoulli's views and is the object of investigation of this book. The corresponding objective Bayesian interpretation of probability construes probabilities as rational degrees of belief, where 'rational' is understood in terms of the three norms: Probability, Calibration, and Equivocation.

In terms of the desiderata of §2.1, the various Bayesian interpretations chiefly disagree with respect to Objectivity. According to the strictly subjective Bayesian interpretation, probability is largely a matter of personal choice; thanks to the Probability norm, the axioms of probability ward off inconsistency, but one can otherwise believe what one likes. Thus, you are perfectly rational if you strongly believe that the moon is made of blue cheese, provided you strongly *disbelieve* that it is not the case that the moon is made of blue cheese. This laxity is often considered to be a stumbling point for the strictly subjective interpretation. Although our current evidence may not logically contradict the proposition that the moon is made of blue cheese, it does, apparently, make it rather improbable; yet the strictly subjective theory cannot account for this. A move to the empirically based subjective Bayesian interpretation helps here: plausibly, current astronomical evidence suggests that a heavenly object has low physical probability of being made of blue cheese; if so, we should scarcely believe that moon is made of blue cheese. While empirically based subjectivism fares better with respect to Objectivity, it suffers from an analogous problem that arises where evidence is currently lacking. Suppose I tell you that I have performed an experiment that has two possible outcomes, positive or negative. To what extent do you believe that the outcome was positive? According to empirically based subjective Bayesianism you may have any degree of belief you wish—for you have no evidence bearing on the physical probability of a positive outcome. Yet this is just as counter-intuitive as condoning a belief that the moon is made of blue cheese: it seems foolhardy to fully believe that the outcome was positive if you have no evidence that points in this direction. Objective Bayesianism gets round this problem by insisting that you should equivocate between possibilities in the absence of evidence: you should believe the outcome was positive to degree $\frac{1}{2}$ (and hence that it was negative to degree $\frac{1}{2}$).

The Bayesian interpretations of probability also disagree as to the epistemology of probability. According to strict subjectivism, probabilities are determined by introspection or by observing behavioural consequences of belief, such as betting behaviour. Empirically based subjectivism admits another route to probability: empirical evidence of frequencies also helps determine probabilities. Finally, the objective Bayesian interpretation admits equivocation as a third determinant of probabilities: evidential symmetry also constrains probability. Apparently, then, the objective Bayesian interpretation admits the most routes to probability, and one might think that it fares best with respect to the Epistemology desideratum because it best explains our practice of determining probabilities via these multifarious routes. The situation is not quite that straightforward, however. Objective Bayesianism also appears to rule out certain routes to probability. In particular, in cases where one's evidence fully determines the strengths of one's beliefs, there is little room for introspection. What matters is not the extent to which one *does* believe propositions of interest, which may be determined by introspection, but rather the extent to which one *should* believe various propositions. This last may be best determined by making the evidence explicit, determining the constraints this evidence imposes on one's degrees of belief, and then calculating which probability function, from all those that satisfy these constraints, is most equivocal. In short, by calculation rather than by introspection or elicitation. Hence, it seems that the subjective approach can account for one more route to probability than can objective Bayesianism.

But this conclusion is too quick. It may not be possible to make all evidence explicit and calculate optimal degrees of belief: we may have too much evidence, we may not be clear about what the evidence is or what constraints it imposes on degrees of belief, or the calculation may be too hard. Then, the best estimate of what one ought to believe may simply be what one does in fact believe. We are, after all, moderately successful in our dealings with the world and presumably that is because the strengths of our beliefs are more or less appropriate. Hence, introspection remains a possibility. More importantly, perhaps, when deciding how strongly to believe a proposition—for example, that it will rain tomorrow—we often elicit opinions from experts: those who have better evidence than we do, or who can better articulate this evidence, or who can better determine the constraints this evidence imposes on degrees of belief, or who can better calculate than we can. This remains possible on the objective Bayesian account: objective Bayesianism by no means precludes resorting to better evidence (§8.3); if that evidence is held by someone else, then it is typically even harder to articulate it oneself and to calculate appropriate probabilities. Then, the only option is to elicit the expert's evidence and reasoning, or, failing that, to elicit the strengths of their beliefs. In sum, then, introspection and elicitation remain possible on an objective Bayesian account, in accordance with the Epistemology desideratum. Having said that, this book will focus on determining probabilities by calculation, rather than by introspection or elicitation which are both well covered in the psychology and the expert systems literature (see, e.g. Oaksford and Chater 2007; Shafer and Pearl 1990).

In terms of Calculi and Parsimony, the three Bayesian interpretations of probability stand or fall together. Regarding the Calculi desideratum, we see in detail in §3.2 that Bayesian epistemology requires that degrees of belief be probabilities (the Probability norm); hence, under a Bayesian interpretation of probability, the probability calculus proceeds as normal.[3] With respect to Parsimony, presumably any ontology will need to be able to provide truthmakers for assertions about rational belief; hence, under a Bayesian interpretation, probabilities require nothing new to be added to the ontology, and the Bayesian interpretations are parsimonious. Are they equally parsimonious? Empirically based subjective Bayesianism and objective Bayesianism both appeal to the notion of physical probability (the varieties of which will be described below). One might think that this would incur an extra ontological burden. Not necessarily, since the appeal to physical probability is in principle eliminable (§9.3): all that is required is that *evidence of* physical probabilities (e.g. sample frequencies, well-supported theoretical claims about chances, theoretically derived symmetries) should constrain degrees of belief, not that these probabilities exist themselves—they might be treated as convenient fictions.

Exactly which stance is taken here can affect how these interpretations fare with respect to the Variety desideratum. When discussing Bernoulli we saw that the Bayesian interpretations naturally handle claims about single cases—probabilities can be attached to anything that can be construed or reformulated as a proposition, but not directly to generic or repeatably instantiatable

---

[3]Some Bayesians argue that Countable Additivity should be replaced by Finite Additivity, which requires additivity over finite partitions (see, e.g. Howson and Urbach 1989*b*, §2.d). While Finite Additivity may fit well with a logical view of probability (Howson 2009), it is arguably not enough for a Bayesian view since an agent whose degrees of belief are finitely but not countably additive are irrational in the sense that they are open to Dutch book—see Theorem 9.1.

attributes—yet one can seemingly sensibly say, for example, that the probability that a (generic or arbitrary) man in the US will develop prostate cancer is about a fifth. Empirically based subjective Bayesian interpretations and objective Bayesian interpretations that are genuinely pluralist about probability, invoking the notion of generic, physical probability, can, of course, handle such claims by appealing to the physical interpretation. But Variety is salvaged at the expense of Parsimony; this route is not open to the fictionalist about physical probability. The only other option is to somehow reconstruct the notion of generic probability from within the Bayesian interpretation. De Finetti tried this approach by appealing to the notion of *exchangeability*, a kind of equivocation condition: given a sequence of outcomes, positive or negative, an agent's prior degrees of belief are *exchangeable* iff for each number m, the agent assigns the same degree of belief to each possible sequence of outcomes that yields m positive outcomes in total. If degrees of belief satisfy this formal exchangeability condition then generic frequencies can be construed as the result of a long sequence of updates by Bayesian conditionalization (de Finetti 1937). The hitch is that under de Finetti's strict subjectivism, there is no reason to suppose that degrees of belief will be exchangeable. The other flavours of Bayesianism can capture generic probabilities by treating them as particular degrees of belief: if all you knew about Bill is that he is a man in the United States, and you had good evidence about the incidence of prostate cancer, and your degree of belief that Bill will get prostate cancer ought to be about a fifth, then the generic probability of a man in the United States getting prostate cancer is about a fifth. Thus, it is possible for the empirically-based subjective Bayesian and the objective Bayesian interpretations to have their cake and eat it: they can be fictionalist about physical probability yet capture the idea of generic probabilities.

2.2.3 *Logical probability*

Laplace's view of probability retained the three key elements which were introduced by Bernoulli and which form the basis of the objective Bayesian interpretation: (i) probabilities are interpreted as degrees of belief, (ii) they are constrained by evidence where available, and (iii) in the absence of such evidence they equivocate between basic alternatives.

> The curve described by a simple molecule of air or vapour is regulated in a manner just as certain as the planetary orbits; the only difference between them is that which comes from our ignorance.
>
> Probability is relative, in part to this ignorance, in part to our knowledge. We know that of three or a greater number of events a single one ought to occur; but nothing induces us to believe that one of them will occur rather than the others. In this state of indecision it is impossible for us to announce their occurrence with certainty. It is, however, probable that one of these events, chosen at will, will not occur because we see several cases equally possible which excludes its occurrence, while only a single one favours it.
>
> The theory of chance consists in reducing all the events of the same kind to a certain number of cases equally possible, that is to say, to such as we may be equally

undecided about in regard to their existence, and in determining the number of cases favorable to the event whose probability is sought. The ratio of this number to that of all the cases possible is the measure of this probability, which is thus simply a fraction whose numerator is the number of favorable cases and whose denominator is the number of all cases possible. (Laplace 1814, pp. 6–7)

As Laplace acknowledges

But that supposes the various cases equally possible. If they are not so, we will determine first their respective possibilities, whose exact appreciation is one of the most delicate points of the theory of chance. (Laplace 1814, p. 11)

Alas, the delicate nature of this point leads Laplace to largely ignore it. He thus focuses on considerations of equivocation at the expense of shedding light on how probabilities are determined when evidence is such that there is no natural partition of events about which one is equally undecided.

This focus on equivocation at the expense of calibration with evidence is a characteristic of the *logical interpretation of probability*. W.E. Johnson was an influential proponent of this view; his thoughts were published posthumously as Johnson (1932a,c,b). John Maynard Keynes developed the logical view in some detail in Keynes (1921). According to Keynes, probability is to be interpreted as a relation between propositions that generalizes the entailment relation of deductive logic: if θ deductively entails φ, then $P(\varphi|\theta) = 1$; if θ deductively entails $\neg\varphi$, then $P(\varphi|\theta) = 0$; otherwise, $P(\varphi|\theta)$ takes some objectively determined value in $[0, 1]$. Assuming one knows just θ, one then ought to believe φ to the extent of this logical probability $P(\varphi|\theta)$. Hence rational degrees of belief are probabilities:

In its most fundamental sense, I think, [probability] refers to the logical relation between two sets of propositions .... Derivative from this sense we have the sense in which ... the term *probable* is applied to degrees of rational belief. (Keynes 1921, pp. 11–12)

Keynes argued that equivocation should be cashed out in terms of the *Principle of Indifference*, also known as Laplace's *Principle of Insufficient Reason*:

The principle of indifference asserts that if there is no *known* reason for predicating of our subject one rather than another of several alternatives, then relatively to such knowledge the assertions of each of these alternatives have an *equal* probability. Thus *equal* probabilities must be assigned to each of several arguments, if there is an absence of positive ground for assigning *unequal* ones. (Keynes 1921, p. 45)

Where this principle is not applicable, Keynes held that probability does not have a precise numerical value (except in extreme cases where the laws of logic and probability suffice to determine probability values). The laws of probability may imply that one probability is greater than another (Keynes 1921, chapter 5), but in the absence of precise numerical values Keynes' account falls well short of telling us exactly how strongly to believe a proposition.

Keynes' logical account can be assessed by appealing to the desiderata of §2.1. The objectivity of probability is assured by taking probability to be first and foremost a logical relation between propositions (and only derivatively a measure of rational degree of belief). The Calculi desideratum is only partially met: logical probabilities are not always numerical so are not probabilities in the usual sense of the word; however, where they are numerical they do satisfy the axioms of probability, and where they only admit qualitative comparison they are constrained by the axioms. From the point of view of Epistemology, the Principle of Indifference tells us almost everything there is to know about logical probabilities. This is a point in favour of Keynes' theory, in that logical probabilities turn out to be knowable, but a point against inasmuch as the logical theory fails to account for other apparent ways of determining probabilities. Turning to Variety, Keynes' theory only interprets single-case probabilities since propositions are the relata of the probability relation. It does not interpret generic probabilities: Keynes argues that a frequency theory might interpret such probabilities, but that any frequency theory faces insurmountable problems (Keynes 1921, chapter 8). Consequently, Keynes' approach only partially interprets probability. Regarding Parsimony, Keynes takes logical probability to be a notion different from degree of belief; hence, an ontology must be rich enough to include such logical relations over and above rational degrees of belief.

Logical theories of probability were also developed by Harold Jeffreys (1931, 1939) and Rudolf Carnap (1950, 1952). Both authors construed logical probability as being fully numerical, thus meeting the Calculi desideratum. Now, if probability is to be a fully numerical logical relation between propositions, then it must be the case that every probability value $P(\varphi|\theta)$ is uniquely determined by θ and φ. But it is not clear how or why this uniqueness condition can be satisfied.

There are two problems here. First, the Principle of Indifference requires a partition of basic alternatives over which to equivocate; this partition is not determined by two propositions θ and φ on their own. In the absence of such a partition the Principle of Indifference cannot be used to specify, for instance, the probability that the moon is made of Stilton given that it is made of blue cheese. If not the Principle of Indifference, then what does determine this probability? As we see, an objective Bayesian interpretation, in contrast with a logical interpretation, carries with it a natural partition over which to equivocate, namely the finest-grained partition of possibilities describable within an agent's language. So the problem disappears if probabilities are construed as relative to agent as well as proposition—a move that distinguishes objective Bayesianism from the logical interpretation of probability.

The second problem arises in the case in which there is an appropriate partition of basic alternatives, but one that is infinite. If there is a countably infinite partition, then, even in the absence of any information discriminating between members of the partition, the Principle of Indifference cannot be applied (see, e.g. Williamson 1999). In order for Countable Additivity to hold, some member of the partition must be given greater probability than some other (for, if all members had the same probability, then if that probability were zero, there would be zero probability that some member of the partition should turn out true; on the other hand, if all members had the same non-zero probability, the probability that some member of the partition should turn out to be true would be infinite;

but by Countable Additivity this probability is 1). Now, in the presence of total indifference there is no grounds for saying of any two members of the partition which one should be ascribed greater probability. Hence, the probabilities over this partition are not uniquely determined. The only way round this problem is to drop the requirement that probabilities are always uniquely determined. This move is at odds with the logical interpretation but quite acceptable to the objective Bayesian.

The logical interpretation of probability and the objective Bayesian interpretation are often conflated or thought of as being very closely related (see, e.g. Franklin 2001; Rowbottom 2008). But as we have seen there are several important differences between the two interpretations. The logical interpretation considers probability to be fundamentally a logical relation that only indirectly concerns degree of belief while the objective Bayesianism interprets probability directly in terms of degree of belief. The logical interpretation typically focuses on equivocation at the expense of calibration, while the objective Bayesian interpretation takes the Equivocation norm to be subsidiary to the Calibration norm—one should only equivocate to the extent that calibration with evidence does not fully determine which degrees of belief to adopt. The logical interpretation takes the logical relation of probability to be relative just to the propositions it relates, while the objective Bayesian interpretation takes probability to be relative to an agent—specifically, to an agent's evidence and language in a particular operating context. A logical interpretation either takes some probabilities to have no numerical value (in which case it suffers with respect to the Calculi desideratum) or it takes probability to be fully numerical (in which case the uniqueness of probability required by the logical interpretation becomes implausible). On the other hand, the objective Bayesian interpretation takes probability to be fully numerical but does not insist on uniqueness.

2.2.4 *Physical probability*

If logical theories of probability focus on equivocation at the expense of evidence of physical probability, physical theories of probability do the opposite.

Physical theories hold that probability is a physical quantity, rather like mass or charge. There are three kinds of physical theory of probability: frequency theories, chance theories, and propensity theories. Frequency theories maintain that probability is generic, a property of an (often infinite) sequence of attributes of individuals (see, e.g. Venn 1866; von Mises 1928; Reichenbach 1935). The frequency of an attribute is the proportion of places in the sequence at which the attribute occurs. For chance theories, probability is single case, with chance typically attaching to an event and indexed by time (see, e.g. Lewis 1980). Propensity theories may either be generic or single case: a propensity to yield a certain frequency distribution may, for example, be attributed to a type of experiment, or a state of the universe (Gillies 2000*b*).

Returning to the desiderata of §2.1, we see that all these physical theories fare well with respect to Objectivity: while physical probability is relative to sequence, time, or state, once these contextual factors are fixed there is no room for disagreement about probabilities. Physical theories do less well with respect to the Calculi desideratum. Countable Additivity can fail for a limiting frequency theory: let E_i be the attribute that *has engine number i* and consider an infinite sequence of observations of engine numbers of cars; each attribute will have limiting frequency 0, and hence $\sum_{i=1}^{\infty} P(E_i) = 0$. One might *stipulate* that chance or propensity satisfies Countable Additivity, but typically there is no way of demonstrating this, and the opponent of such a theory can simply claim the opposite. Turning to the Epistemology desideratum, frequency and propensity theories do well in explaining how we can learn probabilities from evidence of frequencies, while chance theories cope well with symmetry arguments and probabilities posited in scientific theories. None of these theories hold that indifference in the absence of evidence tells us anything about probabilities.[4] While each account covers some of our uses of probability, no single account makes sense of them all, hence the Variety desideratum leaves much to be desired.

One version of the frequency account does well by Parsimony: there is no problem at all in positing *actual frequencies*—frequencies defined with respect to a finite sequence of attributes of actual individuals within some reference class—since nothing is posited beyond the individuals themselves. Indeed, throughout this book we take actual frequencies for granted, and we take it for granted that *sample frequencies*—actual frequencies in a sequence of observations—provide good evidence about probabilities. (This is not to say that the actual frequency account provides an acceptable interpretation of probability; other desiderata may not be considered to be adequately met.) A second version of the frequency account is not so parsimonious: *counterfactual frequencies*—frequencies which would be obtained in the limit were the sequence of attributes of actual individuals in the chosen reference class to be extended indefinitely—require a rich ontology capable of supplying truthmakers for assertions about these infinite sequences. As Venn puts it

> What, for instance, is the meaning of the statement that two children in three fail to attain the age of sixty-three? It certainly does not declare that in any given batch of, say thirty, we shall find just twenty that fail: whatever might be the strict meaning of the words, this is not the import of the statement. It rather contemplates our examination of a large number, of a long succession of instances, and states that in such a succession we shall find a numerical proportion, not indeed fixed and accurate at first, but which tends in the long run to become so. (Venn 1866, pp. 4–5)

> We say that a certain proportion begins to prevail in the long run; but then on looking closer at the facts we find that we have to express ourselves hypothetically, and to say that if present circumstances remain as they are, the long run will show its characteristics without disturbance. When, as is often the case, we know nothing accurately of the circumstances by which the succession of events is brought about, but have strong reasons to suspect that these circumstances are likely to undergo

---

[4]The theory of von Kries (1886), which bridges the logical and the physical approaches, tries to eradicate the Principle of Indifference as outlined earlier in favour of a principle that sets equal probabilities in the face of knowledge of symmetry. Logical probabilities are replaced by chances.

some change, there is really nothing else to be done. We can only introduce the conception of a limit, towards which the numbers are tending, by assuming that these circumstances do not change; in other words, by substituting a series with a fixed uniformity for the actual one with the varying uniformity. (Venn 1866, p. 18)

Several chance and propensity theories posit probabilities as primitive physical quantities, also falling foul of Parsimony. One might simply bite the bullet here and accept that scientific theories need to posit chance in the same way that they need to posit charge. Alternatively, one might seek instead to analyse chance in terms of patterns across possible worlds (Lewis 1994)—but this move does not appeal to those who are not realist about possible worlds and who cannot find in the actual world the truthmakers of claims about possible worlds. Yet another option is to analyse chance in terms of *ultimate beliefs*, that is, interpret chances as the degrees of belief one ought to adopt were one to have total evidence about the world (Lewis 1980, pp. 98–100). This option, discussed further in §9.3, may well be viable, but it takes us away from the realm of physical interpretations of probability.

For ease of exposition, throughout this book we appeal to the physical chance (represented by the probability function P^*). But we see in §9.3 that this apparently unparsimonious appeal to chance is in fact eliminable in favour of talk of indicators of chances such as sample frequencies and symmetries.

2.2.5 The renaissance of objective Bayesianism

We have seen that Bernoulli, Bayes, and Laplace grasped the three tenets that underlie objective Bayesian epistemology: degrees of belief should be probabilities (the Probability norm), calibrated with evidence of physical probability (the Calibration norm), and otherwise equivocal (the Equivocation norm). During the subsequent dark ages, probabilists were led astray by too narrowly focusing on one of these aspects at the expense of the others. Thus, strict subjectivism focused on the axioms of probability, empirically based subjectivism and physical theories on physical probabilities, and logical theories on equivocation.

It was not until the second half of the twentieth century that the three norms were reunited in what is now known as objective Bayesianism. New developments arose in physics, with the publication of Jaynes (1957a,b), papers that brought together hitherto seemingly different uses of the entropy equation, $H(P) = -\sum_{\omega \in \Omega} P(\omega) \log P(\omega)$, in statistical mechanics, and in information theory. Jaynes puts his approach thus:

> This problem of specification of probabilities in cases where little or no information is available, is as old as the theory of probability. Laplace's 'Principle of Insufficient Reason' was an attempt to supply a criterion of choice, in which one said that two events are to be assigned equal probabilities if there is no reason to think otherwise. However, except in cases where there is an evident element of symmetry that clearly renders the events 'equally possible', this assumption may appear just as arbitrary as any other that might be made. Furthermore, it has been very fertile in generating paradoxes in the case of continuously variable random quantities, since intuitive notions of 'equally possible' are altered by a change of variables. Since the time of Laplace,

this way of formulating problems has been largely abandoned, owing to the lack of any constructive principle which would give us a reason for preferring one probability distribution over another in cases where both agree equally well with the available information.

... our problem is that of finding a probability assignment which avoids bias, while agreeing with whatever information is given. The great advance provided by information theory lies in the discovery that there is a unique, unambiguous criterion for the 'amount of uncertainty' represented by a discrete probability distribution. (Jaynes 1957a, p. 622)

Information theory appeals to entropy to measure the amount of uncertainty represented by a probability function. This led Jaynes to argue that the probability function which is least biased, from all those that fit the given information, is the probability function with maximum entropy, and to formulate the *Maximum Entropy Principle* (or *maxent* for short): one should ensure that one's degrees of belief conform to the probability function, from those that fit the given information, that has maximum entropy:

The principle of maximum entropy may be regarded as an extension of the principle of insufficient reason (to which it reduces in case no information is given except enumeration of the possibilities $[\omega]$), with the following essential difference. The maximum-entropy distribution may be asserted for the positive reason that it is uniquely determined as the one which is maximally noncommittal with regard to missing information, instead of the negative one that there was no reason to think otherwise. Thus the concept of entropy supplies the missing criterion of choice which Laplace needed to remove the apparent arbitrariness of the principle of insufficient reason, and in addition it shows precisely how this principle is to be modified in case there are reasons for 'thinking otherwise'. (Jaynes 1957a, p. 623)

Roger Rosenkrantz called the resulting position 'objective Bayesianism':

To uphold an objective Bayesian position, one must extend Laplace's form of the Principle of Insufficient Reason to a method of objectively representing any prior information. Taking our clue from Newton's Principle of Conservative Extrapolation, and following E.T. Jaynes, we arrive at the *maximum entropy rule: choose that prior [probability function] which, among all those consistent with the prior information, maximizes uncertainty.* When prior information assumes the form of constraints governing the relevant random variable, we maximize entropy subject to the constraints. (Rosenkrantz 1977, pp. 54–5)

Rosenkrantz' terminology has now become standard, and this book adopts a similar characterization of objective Bayesianism: objective Bayesian epistemology is any position which holds that the strengths of one's beliefs should be representable by a probability function, from all those that satisfy constraints imposed by evidence, that is maximally equivocal.[5] The version of objective

---

[5]To avoid confusion, it is worth pointing out that Wesley Salmon (2005, p. 150) used the terminology 'objective Bayesianism' in a non-standard way. Salmon's 'objective Bayesianism' is in fact closer to the frequency theory and to empirically based subjective Bayesianism.

Bayesianism explored in this book in many cases is implementable using the Maximum Entropy Principle. (Unlike some advocates of objective Bayesianism, however, we do not assume that agent's evidence always uniquely determines her belief function; neither will we insist that her belief function be updated by Bayesian conditionalization.)

Jaynes' methods were soon in the textbooks. For example, Tribus (1961) developed statistical mechanics from an objective Bayesian point of view:

> In 1957, E.T. Jaynes of Stanford University published a paper in which he showed, starting with Shannon's results, how one could obtain in an elementary way all the results given by Gibbs. It is difficult to overestimate the value to scientific education of this contribution. At one fell swoop the enormous conceptual difficulties inherent in Gibbs' approach have been swept away. It now becomes possible to begin by considering the properties of the smallest particles and by mathematical methods to deduce the properties of large systems. The results obtained by Joule, Mayer, Carnot, Maxwell, Boltzmann and Gibbs are seen in a new light. It is no longer necessary to cover the development of thermodynamics in the same way as the subject developed in history. New generations of students can, in shorter time, reach a state of understanding heretofore impossible for more than a small number of persons. It is this purpose that prompted the writing of this book. (Tribus 1961, p. xx)

The influence of maxent spread. Tribus (1969) presented a detailed account of rational inference and decision from an objective Bayesian point of view, and applied this formalism to making decisions about engineering design. Hobson (1971) further developed the application of maxent to statistical mechanics. Moreover, Rosenkrantz (1977) developed an objective Bayesian philosophy of science. On 2–4 May 1978, there was a conference on *The Maximum Entropy Formalism* at the Massachusetts Institute of Technology; applications of maxent in physics featured prominently, but there were also applications to biology and a range of theoretical developments of the formalism (Levine and Tribus 1979). This was but one of an ongoing series of conferences on maxent and its applications. By the 1980s, objective Bayesian and maxent methods were applied in a wide variety of sciences, as well as in methodological disciplines such as artificial intelligence, statistics, and philosophy. Today, while these methods continue to be applied within physics, they can also be found in fields as diverse as computational linguistics, economic history, logistics, and clinical trials.

2.3 A characterization of objective Bayesianism

In this section, we will be more precise about what we take objective Bayesianism to be.

As discussed in Chapter 1, this book is primarily concerned with objective Bayesian epistemology, that is, the question of how strongly an agent should believe the various propositions expressible in her language. In order to be precise, we assume in this section that the agent's language $\mathcal{L}$ can express n elementary propositions; they are elementary in the sense that they can be expressed by sentences $A_1, \ldots, A_n$ that are not themselves composed of other sentences by means of the usual connectives *not, and, or, implies,* and *iff*. Thus, $\mathcal{L}$ can be thought of as a finite propositional language with propositional variables $A_1, \ldots, A_n$, and where logically complex sentences can be constructed using the usual connectives $\neg, \wedge, \vee, \rightarrow,$ and $\leftrightarrow$. $S\mathcal{L}$ denotes the set of sentences of $\mathcal{L}$. We defer consideration of richer languages to Chapter 5 and §9.1, since many of the key points we wish to make about objective Bayesianism can be made in the more perspicuous propositional framework. As explained in §1.4, we do not assume that the agent's total evidence $\mathcal{E}$ is expressible in her language $\mathcal{L}$.

We take objective Bayesianism to impose three norms on the assignment of the strengths of one's beliefs: Probability, Calibration, and Equivocation. These norms can be explicated as follows.

2.3.1 *Probability*

The Probability norm can be expressed thus:

Probability: The strengths of an agent's beliefs should be representable by a probability function.

We assume that our notional agent is rational in the sense that she complies with the norms of objective Bayesianism; we denote the probability function on $\mathcal{L}$ that represents the strengths of the agent's beliefs by $P_{\mathcal{E}}^{\mathcal{L}}$, or more simply by $P_{\mathcal{E}}$ when $\mathcal{L}$ is fixed, and call it the agent's *belief function*. In §2.1 we encountered the standard definition of a probability function over a field of sets, but we now need to say what we mean by a probability function over a finite propositional language $\mathcal{L}$. Define an *elementary outcome* or *atomic state* ω to be a sentence of the form $\pm A_1 \wedge \cdots \wedge \pm A_n$, where $+A_i$ is just A_i and $-A_i$ is $\neg A_i$. Ω is the set of all 2^n atomic states. A *probability function* on $\mathcal{L}$ is a function P, from sentences of $\mathcal{L}$ to real numbers, that satisfies the properties

P1: $P(\omega) \geq 0$, for each $\omega \in \Omega$,

P2: $P(\tau) = 1$, for some tautology $\tau \in S\mathcal{L}$, and

P3: $P(\theta) = \sum_{\omega \models \theta} P(\omega)$, for each $\theta \in S\mathcal{L}$.

Axioms P1 and P2 set the scale: probabilities have values between 0 and 1. P3 implies that the probabilities of the atomic states are basic: all other probabilities can be defined in terms of them. This axiomatization coincides with the definition of §2.1 if we take Ω to be the outcome space and, setting $\bar{\theta} \stackrel{\mathrm{df}}{=} \{\omega \in \Omega : \omega \models \theta\}$, for each sentence $\theta \in S\mathcal{L}$, we take $\mathcal{F} = \{\bar{\theta} : \theta \in S\mathcal{L}\}$ to be the event space. We use $\mathbb{P}^{\mathcal{L}}$, or, if $\mathcal{L}$ is fixed, simply $\mathbb{P}$, to denote the set of probability functions on $\mathcal{L}$. The Probability norm then says that $P_{\mathcal{E}}^{\mathcal{L}} \in \mathbb{P}^{\mathcal{L}}$.

Given a probability function P on $\mathcal{L}$, we define conditional probability by $P(\theta|\varphi) = P(\theta \wedge \varphi)/P(\varphi)$, for any sentence θ of $\mathcal{L}$ and any sentence φ of $\mathcal{L}$ such that $P(\varphi) > 0$.

As explained in §3.2, the Probability norm is typically justified by betting considerations: degrees of belief are indicative of betting intentions, and if the agent is to avoid bets that lose money whatever happens, her degrees of belief had better behave like probabilities.

2.3.2 Calibration

We take the calibration norm to be

Calibration: The agent's degrees of belief should satisfy constraints imposed by her evidence.

We suppose that there is some set $\mathbb{E} \subseteq \mathbb{P}^{\mathcal{L}}$ of probability functions on $\mathcal{L}$ that are compatible with evidence $\mathcal{E}$ and that the probability function $P_{\mathcal{E}}$ representing the rational agent's degrees of belief lies in that set. Exactly how evidence $\mathcal{E}$ determines the set $\mathbb{E}$ of probability functions is a matter of some debate. In this book, we focus on the following explication of the Calibration norm, which will be explained in more detail in §3.3.

First, if the evidence implies that the physical chance function P^* on $\mathcal{L}$ lies in some set $\mathbb{P}^*$ of probability functions on $\mathcal{L}$, then the agent's belief function $P_{\mathcal{E}}$ should lie in the convex hull $\langle \mathbb{P}^* \rangle$ of $\mathbb{P}^*$, the smallest set containing $\mathbb{P}^*$ that is closed under convex combinations $\lambda P + (1-\lambda)Q$, for $\lambda \in [0,1]$.

Second, qualitative evidence of, for example, causal, logical, hierarchical, or ontological structure imposes certain structural constraints which force $P_{\mathcal{E}}$ to lie in a set $\mathbb{S}$ of probability functions on $\mathcal{L}$ that satisfy those constraints. Structural constraints take the form $P_{\mathcal{E}}^{\mathcal{L}}|_{\mathcal{L}'} = P_{\mathcal{E}'}^{\mathcal{L}'}$, to be read 'when the agent's language is extended from $\mathcal{L}'$ to $\mathcal{L}$ then her new belief function, when restricted to the old language, should match the old function'. For example, if the agent's language is extended to be able to talk about a new event, and if her new evidence is just that this new event is not a cause of the events that she can already talk about, then this change in language and evidence does not warrant any change in beliefs concerning the other events, and the corresponding structural constraint should be satisfied.

Combining the quantitative and qualitative evidential constraints, we have that $\mathbb{E} = \langle \mathbb{P}^* \rangle \cap \mathbb{S}$. The Calibration norm then says that $P_{\mathcal{E}}^{\mathcal{L}} \in \mathbb{E} = \langle \mathbb{P}^* \rangle \cap \mathbb{S}$. The motivation behind Calibration also hinges on the betting interpretation of degrees of belief: well-calibrated degrees of belief avoid certain loss in the long run (§3.3).

2.3.3 Equivocation

Finally, we have the following norm:

Equivocation: The agent's degrees of belief should otherwise be sufficiently equivocal.

Here a function is equivocal to the extent that it is close to the equivocator function $P_=$ on $\mathcal{L}$, which gives each atomic state the same probability,

$$P_=(\omega) = \frac{1}{2^n}.$$

Distance between probability functions is measured by what is sometimes called *cross–entropy* or *Kullback–Leibler divergence*,

$$d(P,Q) = \sum_{\omega \in \Omega} P(\omega) \log \frac{P(\omega)}{Q(\omega)},$$

where $0 \log 0$ is taken to be 0. Note that this is not a distance measure in the usual mathematical sense because it is not symmetric and does not satisfy the triangle inequality.

The motivation behind Equivocation exploits the fact that belief is a basis for action: more extreme degrees of belief tend to trigger high-risk actions (where there is a lot to lose if the agent misjudges) while equivocal degrees of belief are associated with lower risks; and the agent should only take on risk to the minimum extent warranted by evidence (§3.4). Again this can be given a betting interpretation: the agent should equivocate in order to minimize her worst-case expected loss.

One might hold that an agent's degrees of belief should be representable by a probability function $P_\mathcal{E}$ on $\mathcal{L}$ that is in the set of *maximally equivocal evidentially compatible probability functions* $\downarrow\mathbb{E} \stackrel{df}{=} \{P \in \mathbb{E} : d(P, P_=) \text{ is minimized}\}$, as long as there is such a function (i.e. if $\downarrow\mathbb{E} \neq \emptyset$). Since divergence from the equivocator is minimized just when entropy $H(P) = -\sum_{\omega \in \Omega} P(\omega) \log P(\omega)$ is maximized, this gives

Maximum Entropy Principle: An agent's degrees of belief should be representable by a probability function $P_\mathcal{E} \in \{P \in \mathbb{E} : H(P) \text{ is maximized}\}$, as long as there is such a function.

It turns out that in this simple case of a finite propositional language, if $\downarrow\mathbb{E}$ is non-empty, then it is a singleton: $P_\mathcal{E}$ is uniquely determined. This is because entropy is a strictly concave function and it is being maximized over a convex set of probability functions $\mathbb{E}$, so it has a unique maximum if it has a maximum at all. In particular, $\downarrow\mathbb{E}$ is non-empty in the case in which $\mathbb{E}$ is closed, so in that case there is guaranteed to be a (unique) maximum. But in general there is no guarantee of uniqueness and this remains true on richer languages: as we see in Chapter 5 and §9.1, richer languages may also leave some room for subjective choice as to how strongly to believe a proposition.

There is no guarantee of uniqueness here because in certain cases $\downarrow\mathbb{E}$ may be empty. For example, consider an agent whose language can express the elementary proposition U, *the toss of the coin will yield a head uppermost*, and whose evidence concerning U is just that the coin is biased in favour of heads. Plausibly then, for $P \in \mathbb{E}, P(U) > 1/2$ ($\mathbb{E}$ is convex but not closed since there is a limit point $P \notin \mathbb{E}$ such that $P(U) = 1/2$), and for any function P in $\mathbb{E}$ there is a function $Q \in \mathbb{E}$ that is closer to the equivocator, $1/2 < Q(U) < P(U)$. In such a case the objective Bayesian cannot say—just on the basis of the Maximum Entropy Principle and the agent's evidence and language—that the agent should adopt a particular probability function, since there is no maximally equivocal evidentially compatible probability function. The best that can be said is a comparative judgement: one belief function is closer to the equivocator than another. As to which probability functions are *sufficiently close* will depend in part on considerations other than the agent's evidence and language. If the agent can—or need—only distinguish probability values to a fixed finite number of decimal places then any probability function that is indistinguishable from any other more equivocal function in $\mathbb{E}$ will be sufficiently close. We take the objective Bayesian protocol to advocate adopting a belief function that is sufficiently close to the equivocator. (Arguably, this should be so whether or not $\downarrow\mathbb{E}$ is empty.) We denote the set of sufficiently close belief functions—that is, the set of *sufficiently equivocal evidentially compatible probability functions*—by $\Downarrow\mathbb{E}$; this set will typically be larger than $\downarrow\mathbb{E}$. In this book, we usually be more concerned with determining the set $\downarrow\mathbb{E}$ of maximally equivocal functions than with determining the set $\Downarrow\mathbb{E}$ of sufficiently equivocal functions, but it should be borne in mind that contextual factors such as computational capacity can render $\Downarrow\mathbb{E}$ of primary interest.

In sum, the Equivocation norm says that $P_\mathcal{E}^\mathcal{L} \in \Downarrow\mathbb{E}$, where $\downarrow\mathbb{E} \subseteq \Downarrow\mathbb{E} \subseteq \mathbb{E}$. By the Calibration norm, $\mathbb{E} = \langle \mathbb{P}^* \rangle \cap \mathbb{S}$. By the Probability norm, $\mathbb{E} \subseteq \mathbb{P}_\mathcal{E}^\mathcal{L}$.

3

Motivation

In §2.3, objective Bayesianism was characterized by appealing to the norms of Probability, Calibration, and Equivocation. In this chapter, we encounter some of the motivation behind this characterization. We see that each norm can be justified by showing that the norm should be followed if degrees of belief are to do what we require of them.

The betting interpretation of belief is introduced in §3.1. Section 3.2 explains the standard betting justification of the Probability norm. Section 3.3 discusses the formulation of the Calibration norm and presents its justification in terms of long-run bets. Section 3.4 presents a justification of the Equivocation norm that appeals to considerations of caution. Finally, §3.5 argues that these epistemological norms are not some voluntary code of conduct, but are forced upon us by the uses to which the strengths of our beliefs are put.

3.1 Beliefs and bets

There are two main ways of arguing for a rational norm N. The first is via an *argument by derivation*. This proceeds by making some assumptions, A and then showing that N follows by the laws of logic, $\models A \rightarrow N$. If A can be chosen to be intuitively plausible, then the thought is that N will be plausible too. This kind of argument is good for convincing someone who has no prior opinion about N and who finds A quite plausible: if the derivation is correct it is clearly hard for someone to accept A but deny N. However, this kind of argument is not very helpful for convincing someone who is already inclined to doubt N and who does not already feel committed to all the assumptions in A. Such a person may reason like this: A is obviously as strong, logically, as N; so if A appears more plausible than N this is sheer artifice—A must have been chosen in such a way as to render the dubious qualities of N innocuous; the derivation shows that $\models \neg N \rightarrow \neg A$; hence, since I doubt N, I should simply doubt A. Since the three norms under consideration here are somewhat controversial, with many detractors as well as proponents, arguments by derivation are far from ideal, unlikely to convince the detractors.

The second main way of arguing for a rational norm is via an *argument by interpretation*. The idea here is to interpret, or explicate, the terms under consideration in a plausible way, and then show that under such an interpretation the norm must hold. This latter task may require some assumptions A', but these assumptions will typically not be doing all the work and may be relatively trivial: the interpretation itself may be providing much of the force of the argument. The argument structure is: under interpretation I, N is construed as saying N', but

$\models A' \to N'$ and assumptions A' are plausible, hence so is N' and thus N. In response, someone who doubts N may question either the assumptions A' or the interpretation I. If, indeed, the assumptions A' are apparently trivial, the question is the interpretation. Of course, it is incumbent upon someone who questions the interpretation of a term to say what it is he is in fact talking about when he uses the term. More than that, though, in order to avoid the charge that he is simply talking at cross-purposes, he should show that his interpretation is a *better* interpretation of the term than the original interpretation I. This is quite possible if I is in some way mistaken, but not an easy thing to do if I remains plausible.

An argument by derivation, then, has the potential to convince someone with no prior opinion on the conclusion in question, but is unlikely to persuade a detractor. On the other hand, an argument by interpretation—though perhaps not entirely conclusive since there is always the possibility that a better interpretation will be found—has the means to persuade the sceptic. The approach here will be to favour arguments by interpretation in the hope that by appeal to judicious interpretations, detractors may be convinced.

The key interpretation that we will appeal to is a betting interpretation of rational belief. Our beliefs guide our actions, and, as Frank Ramsey pointed out, our actions can be thought of as bets:

> all our lives we are in sense betting. Whenever we go to the station we are betting that a train will really run, and if we had not a sufficient degree of belief in this we should decline the bet and stay at home. (Ramsey 1926, p. 183)

Formally, the *betting interpretation of rational belief* proceeds as follows. 'The agent should believe proposition θ to degree q' is interpreted as saying that, were the agent to offer to buy or sell bets on θ at the same price, the price should be qS, where S is the stake bet. That is, the agent should deem qS to be a fair price to have to pay in order to receive S if θ turns out to be true, where $S \in \mathbb{R}$ is an unknown stake (positive or negative) that can depend on q. (q is called the agent's *betting quotient* for θ.) Conversely, if the agent ought to announce betting quotient q for θ, then we can say that she should believe θ to degree q.

This betting set-up extends to multiple propositions—the agent may make multiple simultaneous bets on different propositions—but only within the boundaries of common sense: in particular, we do not permit combinations of bets that could lead to an infinite amount of money changing hands once the bets are settled. This restriction to finite settlements might be achieved by allowing only finitely many bets at any one time, or, if infinitely many simultaneous bets are allowed, by restricting the set of propositions and stakes to ensure that only a finite amount of money can change hands; the details need not delay us here. It is also assumed that the stake-maker has just the same evidence to hand as the agent setting the betting quotients—neither party is given a prior advantage over the other.

Since betting has been appealed to in a variety of ways in the literature on Bayesianism, it is worth being clear about what the betting set-up is doing here.

Note, first, that the betting set-up yields an interpretation of *rational* degree of belief, the degree to which the agent *ought* to believe a proposition. There is no assumption that people's betting quotients always do align with their elicited degrees of belief. Second, note that this is intended as an interpretation rather than a measurement device. Ramsey (1926) suggested that betting behaviour could be used to measure an agent's actual degrees of belief; this suggestion faced a number of objections in the literature (see, e.g. Seidenfeld *et al.* 1990). Here, instead, we use the betting set-up to specify the *meaning* of rational degree of belief. (On the objective Bayesian approach, measurement is far less important than under subjective Bayesianism since rational degrees of belief tend to be calculated—via the Maximum Entropy Principle, for instance—rather than elicited.)

Third, note that this is interpretation in the sense of *explication* (Carnap 1950, chapter 1). The idea is that we make precise a core feature of the imprecise concept in question, rational belief, and we replace the imprecise concept with the explicating concept, betting quotients in our case. The interpretation is to be judged successful if the chosen core feature—in our case, strength of belief as it guides behaviour—is in fact central to our purposes; if any ancillary connotations of the imprecise concept can safely be ignored given our purposes; if the interpretation is fruitful; and if the interpreting concept is simple and precise. In particular, it is not expected that the interpretation be true to *every* use of the interpreted concept. An agent may rationally refuse to gamble at all—perhaps because she has good evidence that gambling leads to ruin, or perhaps because she simply has no funds. Or she may rationally refuse to gamble in the way outlined above, perhaps because she wants to limit her possible loss or will only gamble when she reasonably expects financial gain. But our betting interpretation ignores such confounding situations by considering only the case in which the agent *were* to bet as required. Or, an agent may rationally gamble for the sake of it, not caring about gains or losses—a rich agent's wish for excitement may outweigh any pecuniary repercussions. The above interpretation also sets aside such confounding intentions.

3.2 Probability

There are two standard arguments for the Probability norm, an argument by derivation called *Cox's theorem*, and an argument by interpretation, the *Dutch book argument*.

Cox's theorem aims to show that any belief function that satisfies certain assumptions must be a probability function. The result was originally presented in Cox (1946), but the validity of the argument, as he presented it, is a matter of controversy (Paris 1994, p. 24; Halpern 1999a, 199b; Howson 2009, footnote 18). We shall give the result as formulated and proved by Paris (1994, pp. 24–32); the conditional belief $Bel(\theta|\psi)$ should be thought of as 'some measure of the reasonable credibility of the proposition θ when the proposition ψ is known to be true' (Cox 1946, p. 6, our notation):

Theorem 3.1. (Cox/Paris) *Suppose that a conditional belief function $Bel(\theta|\psi) : S\mathcal{L} \times S\mathcal{L} \longrightarrow [0,1]$ is defined for each consistent ψ and each θ, that $\varphi \wedge \psi$ is consistent, and*

1. *if θ is logically equivalent to θ' and ψ is logically equivalent to ψ', then $Bel(\theta|\psi) = Bel(\theta'|\psi')$,*
2. *if ψ logically implies θ, then $Bel(\theta|\psi) = 1$ and $Bel(\neg\theta|\psi) = 0$,*
3. *there is some continuous function $F : [0,1] \times [0,1] \longrightarrow [0,1]$ which is strictly increasing in both coordinates on $(0,1] \times (0,1]$ such that $Bel(\theta \wedge \varphi|\psi) = F(Bel(\theta|\varphi \wedge \psi), Bel(\varphi|\psi))$,*
4. *there is some decreasing function $S : [0,1] \longrightarrow [0,1]$ such that $Bel(\neg\theta|\psi) = S(Bel(\theta|\psi))$,*
5. *for any $a,b,c \in [0,1]$ and $\epsilon > 0$, there are $\theta_1, \theta_2, \theta_3, \theta_4 \in S\mathcal{L}$ such that $\theta_1 \wedge \theta_2 \wedge \theta_3$ is consistent and $Bel(\theta_4|\theta_1 \wedge \theta_2 \wedge \theta_3), Bel(\theta_3|\theta_1 \wedge \theta_2), Bel(\theta_2|\theta_1)$ are within ϵ of a,b,c, respectively,*

then there is a continuous, strictly increasing, surjective function $g : [0,1] \longrightarrow [0,1]$ such that $g\,Bel(\cdot|\tau)$ is a probability function on $\mathcal{L}$, for any tautology τ.

Thus, $Bel(\cdot|\tau)$ is isomorphic to a probability function on $\mathcal{L}$.

While the assumptions might appear innocuous, and, in the absence of any strong feelings about the conclusion, render it more plausible, it is not hard to find assumptions upon which someone who doubts the conclusion can cast aspersions. The assumption that a conditional belief function maps sentences to unique real numbers is rejected by the imprecise probability approach to belief, which holds that the strength of a belief is better measured by an interval of numbers than by a single number. Assumptions 1 and 2 would trouble anyone who thinks that it is too much to expect an agent to be logically omniscient, that is, always to be able to tell whether two sentences are logically equivalent or whether one logically implies the other. Assumption 3 invokes continuity, which seems to be less a rational constraint than a technical convenience. Assumption 4 is objectionable to the proponent of the Dempster–Shafer theory of belief which holds that the degree to which one believes the negation of a proposition is not determined by the degree to which one believes the proposition itself. Assumption 5, which implies that the range of the belief function is dense in the unit interval and hence that an agent's belief must have uncountably many gradations, is an extremely strong assumption which even many advocates of the Probability norm would find intolerable. Although assumption 5 can be replaced by alternative assumptions (Halpern 1999b), they are hardly more compelling. In short, this argument by derivation, though championed by subjectivists (see, e.g. Howson 2008) and objectivists (see, e.g. Jaynes 2003) alike, is unlikely to convince those who already have reservations about the Probability norm.

The Dutch book argument, on the other hand, is an argument by interpretation (see, e.g. de Finetti 1937; Paris 1994, pp. 19–23; Williamson 1999). It appeals to the betting interpretation of rational belief and a formal result. The formal result is that if an agent avoids the possibility of certain loss when betting—such an eventuality is called a *Dutch book*—then her betting quotients must satisfy the axioms of probability. Since, according to the betting interpretation, one can go back and forth between rational degrees of belief and rational betting quotients, under the assumption that the agent ought to avoid the possibility of sure loss, her degrees of belief should be representable by a probability function. Here, first, is the formal result, a version of what is sometimes known as the Ramsey–de Finetti theorem (Ramsey 1926; de Finetti 1937).

Theorem 3.2 *Define function $P : S\mathcal{L} \longrightarrow \mathbb{R}$ by $P(\theta) =$ the agent's betting quotient for θ. The agent's bets on propositions expressible in $\mathcal{L}$ avoid the possibility of a Dutch book if and only if they satisfy the axioms of probability:*

P1: $P(\omega) \geq 0$, *for each $\omega \in \Omega$,*
P2: $P(\tau) = 1$, *for some tautology $\tau \in S\mathcal{L}$, and*
P3: $P(\theta) = \sum_{\omega \models \theta} P(\omega)$, *for each $\theta \in S\mathcal{L}$.*

Proof: First, we shall show that if the agent's betting quotients are not Dutch-bookable, then they satisfy the axioms of probability.

P1. Suppose $P(\omega) = q < 0$, for some $\omega \in \Omega$; we show that the agent is susceptible to a Dutch book. Then, the agent's loss on a bet on ω is $L_\omega = (q - I_\omega)S_\omega$, where I is the indicator function on $\mathcal{L}$ defined by

$$I_\theta = \begin{cases} 1 & : \quad \theta \text{ is true} \\ 0 & : \quad \text{otherwise} \end{cases}$$

and where S_ω is a stake, positive or negative, that may depend on P. Suppose the stake-maker defines

$$S_\theta = \begin{cases} -1 & : \quad \theta = \omega \\ 0 & : \quad \text{otherwise} \end{cases}.$$

Then, the agent's total loss over all her bets is $L = L_\omega = I_\omega - q \geq -q > 0$. In other words, the agent's total loss is positive, whatever truth value ω turns out to take: a Dutch book.

P2. Suppose $P(\tau) = q < 1$. Setting

$$S_\theta = \begin{cases} -1 & : \quad \theta = \tau \\ 0 & : \quad \text{otherwise} \end{cases},$$

the agent's total loss is $L = I_\tau - q = 1 - q > 0$, since τ is a tautology. Hence, we have a Dutch book.

On the other hand, if $P(\tau) = q > 1$, one can set

$$S_\theta = \begin{cases} 1 & : \quad \theta = \tau \\ 0 & : \quad \text{otherwise} \end{cases},$$

in which case $L = q - 1 > 0$, a Dutch book.

Hence, if $P(\tau) \neq 1$ for tautology τ, then the agent is prone to a Dutch book.

P3. Suppose $P(\theta) < \sum_{\omega \models \theta} P(\omega)$. Set

$$S_\varphi = \begin{cases} -1 & : \varphi = \theta \\ 1 & : \varphi = \omega \models \theta \\ 0 & : \text{otherwise} \end{cases}.$$

The agent's total loss is

$$L = I_\theta - P(\theta) + \sum_{\omega \models \theta}(P(\omega) - I_\omega) = \left(\sum_{\omega \models \theta} P(\omega)\right) - P(\theta)$$

since $I_\theta = \sum_{\omega \models \theta} I_\omega$. But this is positive, so we have a Dutch book.

On the other hand, suppose $P(\theta) > \sum_{\omega \models \theta} P(\omega)$. Set

$$S_\varphi = \begin{cases} 1 & : \varphi = \theta \\ -1 & : \varphi = \omega \models \theta \\ 0 & : \text{otherwise} \end{cases}.$$

The agent's total loss is

$$L = P(\theta) - I_\theta + \sum_{\omega \models \theta}(I_\omega - P(\omega)) = P(\theta) - \sum_{\omega \models \theta} P(\omega) > 0,$$

a Dutch book again.

Next, we show that if the agent's betting quotients at any one time satisfy the axioms of probability then they are not Dutch-bookable.

Let $\omega_1, \ldots, \omega_{2^n}$ run through the atomic states of $\mathcal{L}$. Suppose the agent bets on sentences θ in her language using betting quotients that satisfy the axioms of probability. The agent's loss if ω_i turns out to be true is

$$L_i = \sum_{\omega_i \models \theta}(P(\theta) - 1)S_\theta + \sum_{\omega_i \not\models \theta} P(\theta)S_\theta = \sum_{\theta \in S\mathcal{L}} P(\theta)S_\theta - \sum_{\omega_i \models \theta} S_\theta.$$

That these sums are finite is ensured by our restriction to combinations of bets under which only a finite amount of money can change hands.

The agent's expected loss is thus

$$\sum_{i=1}^{2^n} P(\omega_i)L_i = \left(\sum_{i=1}^{2^n} P(\omega_i)\right) \sum_{\theta \in S\mathcal{L}} P(\theta)S_\theta - \sum_{i=1}^{2^n}\left(P(\omega_i) \sum_{\omega_i \models \theta} S_\theta\right)$$

$$= 1 \sum_{\theta \in S\mathcal{L}} P(\theta)S_\theta - \sum_{i=1}^{2^n} \sum_{\omega_i \models \theta} P(\omega_i)S_\theta$$

$$= \sum_{\theta \in S\mathcal{L}} P(\theta)S_\theta - \sum_{\theta \in S\mathcal{L}} \sum_{\omega_i \models \theta} P(\omega_i)S_\theta$$

$$= \sum_{\theta \in S\mathcal{L}} P(\theta) S_\theta - \sum_{\theta \in S\mathcal{L}} P(\theta) S_\theta$$
$$= 0.$$

Since the expected loss is zero, it cannot be the case that in every eventuality ω_i the loss L_i is positive. Hence, there is no possible Dutch book. ∎

The structure of the Dutch book argument is as follows. First, we interpret rational degrees of belief as rational betting quotients. Then, we invoke Theorem 3.2 to show that, under the assumption that betting quotients should avoid sure loss, they must satisfy the axioms of probability. We conclude that rational degrees of belief satisfy the axioms of probability. Note that the betting interpretation is doing a lot of work here, and that the assumption (avoiding sure loss) is correspondingly benign.

Someone who doubts the Probability norm has two options if he is to continue to doubt it. He can either question the interpretation or question the assumption. If he questions the interpretation, then it is incumbent upon him to say what he means by rational belief. There are two main rival explications of rational degree of belief. The first is to abandon the idea of monetary bets and to think in terms of utilities (Ramsey 1926; Savage 1954; Skyrms 1987). While this is a more sophisticated explication, it still leads to the Probability norm.[1] The second rival interpretation involves permitting the agent to choose different betting quotients for a bet on θ and a bet against θ. This interpretation (or its utility-based version) is normally used to argue for the imprecise probability theory of belief: interval-valued rather than real-valued beliefs (Walley 1991). This betting set-up is perhaps more appropriate than that of §3.1 as a means for measuring agents' actual degrees of belief, since in practice we are rarely prepared to buy and sell bets at the same rate. However, there are two main reasons why this modified set-up is far less widely accepted than the original set-up as an interpretation of rational degree of belief, rather than as a measurement device. First, the original set-up is simpler and has more intuitive appeal: while we do, in practice, buy and sell bets at different rates, the rate at which we would be prepared to both buy and sell, if we had to, remains a plausible interpretation of strength of belief. Second, we see in §3.4 that the original set-up better captures the connection between belief and decision: in short, if taking an action is to depend on strength of belief and a suitable threshold for taking that action, then strengths of belief need to be linearly ordered, as they are with the betting interpretation of §3.1. So, while it is possible to interpret rational belief in such a way that the Probability norm will fail, arguably the original betting set-up remains the better interpretation.

The other option for an opponent of the Probability norm is to question the key assumption in the Dutch book argument: that an agent's betting quotients

---

[1] Although it is more sophisticated, it is not necessarily better as an explication of belief—see Nau (2001) for reasons to prefer the monetary betting interpretation.

should avoid the possibility of sure loss. Given situations in which an agent does give the requisite betting quotients, is susceptibility to Dutch book really irrational? There are two related reasons why one might think it is not (Baillie 1973). First, the stake-maker might be on the agent's side, not out to Dutch book her, in which case susceptibility to Dutch book will be harmless. Second, the symmetry of the betting set-up ensures that the possibility of sure loss is avoided just when the possibility of sure *gain* is avoided; yet, it seems *irrational* to rule out the possibility of sure gain. Imagine an agent who knows that her guardian angel will ensure that no trouble befalls her: whenever she makes a set of Dutch-bookable bets the angel will see to it that the stake-maker does not actually Dutch book her. Or, better still, the angel will see to it that stakes are chosen to ensure a certain win, whatever happens. Surely, such an agent is by no means irrational to offer Dutch-bookable betting quotients?

The proponent of the Probability norm can capitulate a bit here by admitting that such a scenario is metaphysically possible and that such an agent would not be bound by the Probability norm. Just as the betting interpretation, being an explication, will not be suitable to all uses of the term 'rational belief', so too the norms of rational belief may not apply to all beliefs. It is enough that rational norms govern typical or usual rational beliefs. Typically, of course, both parties in a betting situation are out to win as much as possible: a benevolent stake-maker or a guardian angel is hardly the norm. Moreover, in this typical case a combination of bets can be foolish or irrational whether or not one's opponent in fact exploits it to one's detriment—the mere possibility is enough. More generally, any degree of belief may be called on to inform several future decisions, not just a single bet, and it is rarely if ever plausible to suppose that a poorly judged belief will never lead to personal loss. So the assumption that betting quotients ought not be Dutch-bookable stands often enough for it to underpin the Probability norm.

To conclude, the Dutch book argument is perhaps more convincing than Cox's theorem as a justification of the Probability norm. To be sure, it is a weaker justification in the sense that it admits the possibility that odd circumstances can override the Probability norm in some cases. But, that is not to diminish the status of the norm itself, which need only apply in normal circumstances.

Note that arguably the Probability norm needs restricting in any case in order to cope with an entirely different problem—the problem of *logical omniscience*. Under the Probability norm, $\theta \models \varphi$ implies $P_\mathcal{E}(\theta) \leq P_\mathcal{E}(\varphi)$. So your degrees of belief ought to track logical consequences. But, for particular θ and φ you may have no clue as to whether one logically implies the other, and θ and φ may be complicated enough that you cannot reasonably be expected to work out whether one logically implies the other. So the most one can rationally demand is that *if you grant that* $\theta \models \varphi$, then $P_\mathcal{E}(\theta) \leq P_\mathcal{E}(\varphi)$. Obviously, this qualification significantly complicates the Probability norm, and in cases where logical consequences are trivial or not in question, one might just as well stick with the Probability norm as formulated above. But for some applications, for example, to mathematics, the qualifications can be important (Corfield 2001).

3.3 Calibration

The view that an agent's belief function should be compatible with her evidence is a view shared by all Bayesians, but there are differences as to how this principle is cashed out. A minimum requirement is captured by the following[2]:

C1: If $\theta \in S\mathcal{L}$ is in the agent's evidence base $\mathcal{E}$ and this evidence $\mathcal{E}$ is consistent, then $P_{\mathcal{E}}(\theta) = 1$.

This principle can be motivated by the following kind of consideration: evidence should at the very least be evident to its holder; but this implies that it should be fully believed. This tallies with our use of evidence in this book. We take evidence to be anything that the agent takes for granted in her current operating context. Should she doubt a proposition θ, that is, should she believe $\neg\theta$ to some extent, she can hardly be said to grant θ. Hence, she should fully believe θ. (Note that this is full belief in a provisional sense: in another situation the agent may no longer grant θ, in which case it will no longer be a part of her evidence base and full belief in θ will no longer be a rational requirement.)

Betting considerations can also be used to support this line of argument. We saw in §3.2 that, arguably, the agent should not place bets that open up the possibility of a Dutch book, that is, loss whatever the outcome. If we take the agent's evidence to constrain the possible outcomes under consideration, then a similar argument can be invoked to motivate C1. If $\theta \in S\mathcal{L}$ is in the agent's evidence base $\mathcal{E}$, then the agent grants that θ is true. We have assumed that the stake-maker has the same evidence as the agent who is setting betting quotients; hence, the stake-maker also grants that θ is true. Now, if θ is true, then $\neg\theta$ is not a possibility (since $\mathcal{E}$ is consistent, $\neg\theta$ is neither granted nor is it a tautology). For bettor and stake-maker alike, 'whatever the outcome' is 'whatever the outcome consistent with θ'. In which case, to avoid sure loss the agent's betting quotient for θ must be 1 (the argument is the same as for P2 in Theorem 3.2). Hence, if θ is granted, then it should be fully believed.

Principles like C1 have been generalized to yield what is sometimes called the *Miller's Principle* (Miller 1966) or the *Principal Principle* (Lewis 1980):

C2: If $P^*(\theta) = x$ is the evidence in $\mathcal{E}$ that is most pertinent to θ, and $\theta \in S\mathcal{L}$, then $P_{\mathcal{E}}(\theta) = x$.

Here P^* is taken to be the chance function—single-case physical probability—at the time at which the agent's degrees of belief are being formulated. Given θ, evidence $\mathcal{E}'$ in $\mathcal{E}$ is the evidence most *pertinent* to θ if it is a minimal set of evidence that screens off all other evidence from θ: that is, if $\{P_{\mathcal{E}}(\theta) : P_{\mathcal{E}} \in \mathbb{E}\} = \{P_{\mathcal{E}'}(\theta) : P_{\mathcal{E}'} \in \mathbb{E}'\}$ and there is no $\mathcal{E}'' \subset \mathcal{E}'$ such that $\{P_{\mathcal{E}}(\theta) : P_{\mathcal{E}} \in \mathbb{E}\} =$

---

[2] Not everyone endorses C1: some Bayesians maintain that only logically necessary propositions should have probability 1. But this view does not sit well with probability theory, which regularly attaches probability 1 (or 0) to propositions that are not logically true (respectively false).

$\{P_{\mathcal{E}''}(\theta) : P_{\mathcal{E}''} \in \mathbb{E}''\}$. Here $\mathbb{E}$ (respectively $\mathbb{E}', \mathbb{E}''$) is the set of probability functions that is compatible with evidence $\mathcal{E}$ (respectively $\mathcal{E}', \mathcal{E}''$).

Leaving aside for the moment the question of exactly when evidence is most pertinent, one can ask what motivates C2. There are two main kinds of justification—as to which is the right one depends how chance itself is construed. According to one view, the defining feature of chance is that it constrains belief via C2. C2 then becomes trivially true: if you know that the chance of θ is x, then, since chance is whatever constrains belief via C2, you should believe θ to a degree x. This view was developed by Lewis (1980, 1994). Lewis takes chances to be posited by probabilistic laws, which are consequences of the deductive system that best captures all matters of fact (best in the sense of offering the best balance between simplicity, strength, and fit). For Lewis, the characteristic feature of chance is that it constrains degrees of belief; hence, the Principal Principle needs no independent argument.

According to another view, the characteristic feature of chance is its connection with *frequency* rather than belief. This view was championed by Kolmogorov, who held that 'One can be practically certain that if the complex of conditions C is repeated a large number of times, n, then if m be the number of occurrences of event A, the ratio m/n will differ very slightly from $P^*(A)$' (Kolmogorov 1933, p. 4, our notation). This is perhaps most naturally explicated as: one should grant that the limiting relative frequency will equal the chance, that is, $P^*(A) = \lim_{n \to \infty} m/n$ should be in $\mathcal{E}$. Such a connection between chance and frequency can be motivated by the following argument. The strong law of large numbers says roughly that if the occurrences of A are identically and independently distributed (iid) according to the chance function P^*, then there is no chance (zero chance) that the chance $P^*(A)$ will differ from the limiting relative frequency $\lim_{n \to \infty} m/n$. Now invoke a principle related to C2 but applying to evidence rather than belief: if one can grant that the chance that θ is true is 1, then one should grant θ itself. (Note that θ may have chance 1 yet in fact be false, so this principle is to be interpreted as a default principle, to be overridden if there is more pertinent evidence that θ is in fact false.) Applying this last principle to the law of large numbers, we have that if one grants that occurrences of A are iid with respect to P^*, then one should grant that $P^*(A) = \lim_{n \to \infty} m/n$.

Given this connection between chance and frequency, one can justify a calibration principle of the form of C2 as follows. Suppose that θ asserts a particular occurrence of A under conditions C and that $P^*(\theta) = x$ is the evidence most pertinent to θ.[3] Then the connection between chance and frequency implies that the agent should grant that the limiting relative frequency of A under C is x, for iid occurrences of A under C. Let θ_1 be θ and let $\theta_2, \theta_3, \ldots$ be sentences asserting other particular iid occurrences of A under C concerning which the agent has the same information as for θ (i.e. for which $P^*(\theta_i) = x$ is the evidence most pertinent

---

[3] In a predicate, language θ might take the form $At \wedge Ct$, for some individual constant t. Predicate languages will be discussed in detail in Chapter 5.

to θ_i). Let us assume to begin with that the agent's degrees of belief in the θ_i are identical, say $P_{\mathcal{E}}(\theta_i) = q$. (This is a special case of the *exchangeability* condition introduced in §2.2, which holds when degrees of belief concerning combinations of outcomes are dependent only on the number of occurrences of each outcome and not on the ordering of the occurrences.) Suppose that $q > x$, where x is the value that the agent should grant is the limiting relative frequency of A under C. That x is the limiting relative frequency means that for any $\epsilon > 0$, there is a natural number N such that for $n \geq N$, $|m/n - x| < \epsilon$. Let $\epsilon = |q - x|$. The agent's loss on betting on $\theta_1, \theta_2, \ldots, \theta_n$ with betting quotient q is $\sum_{i=1}^{n}(q - I_{\theta_i})S_i$, where S_i is the stake on the bet on θ_i. Now, the stake-maker can set each $S_i = 1$ so that the agent's loss is $nq - m > nq - n(x + \epsilon) = nq - nq = 0$, for $n \geq N$. Hence, the agent is sure to lose money in the long run. Similarly if $q < x$. On the other hand, if $q = x$, then the stake-maker cannot choose stakes that ensure the agent loses money—this fact is sometimes called the *law of excluded gambling systems* (see, e.g. Gillies 2000a, Chapter 5; Billingsley 1979, §7). Hence, C2 can apparently be justified by betting considerations, just as can the Probability norm.

Note, we have supposed that the agent has identical degrees of belief in the θ_i. So, strictly speaking, we have only shown that if your degrees of belief are *not* calibrated to chances via C2, then you *must* believe propositions θ_i, which are indistinguishable with respect to current evidence, to different degrees. But to the extent that the latter consequence is intuitively untenable, calibration remains the only option.

In response, one might try to defend the apparently undesirable consequence, by pointing out that where evidence fails to distinguish the θ_i, other considerations may yet do so, thereby warranting differing degrees of belief. For example, if $\varphi_1, \varphi_2, \ldots$ form a countably infinite *partition* of propositions (i.e. they are mutually exclusive and exhaustive), then the agent had better believe some φ_i to a greater extent than another φ_j, even if her evidence fails to distinguish φ_i from φ_j. This is because her degrees of belief ought to satisfy Countable Additivity on pain of Dutch book (Theorem 9.1). Then $P_{\mathcal{E}}(\varphi_1) + P_{\mathcal{E}}(\varphi_2) + \cdots = 1$, forcing a non-uniform distribution of belief over the partition (Williamson 1999).

Although this response does indeed point to cases in which a uniform distribution is untenable, such is not the case here. There is no partition here—rather a sequence of iid outcomes—and it is hard to see how non-evidential considerations could force a skewed distribution over the θ_i in this case. So Calibration remains the viable option.

In sum, then, a case can be made for a Calibration norm on either account of chance. If chance is essentially linked to belief, then C2 holds trivially. On the other hand, if chance is essentially linked to frequency, then C2 is the only option if arbitrarily skewed degrees of belief are to be avoided.

Before moving on, it is worth mentioning another kind of justification of C2. The previous justification can be put like this: in order to avoid sure loss in the long run yet permit uniform degrees of belief, C2 should hold. But chance is meant to govern the outcome of events in the short run as well as the long run, and good calibration can be expected to pay off in the short run. Suppose that $P^*(\theta) = x$ is the evidence most pertinent to θ, but that $P_\mathcal{E}(\theta) = q > x$ say. The agent's loss on this bet is as usual $(q - I_\theta)S$. The expected loss is $x(q-1)S + (1-x)qS = (q-x)S$. Now, the stake-maker can choose $S > 0$ to ensure that the expected loss is positive. Similarly, if $q < x$ the stake-maker can choose $S < 0$ to ensure that the expected loss is positive. But if $q = x$, then the expected loss is zero. Intuitively, since it is within the agent's power to avoid positive expected loss, she should avoid it. But doing so requires setting degrees of belief to chances via C2.

Of course, positive expected loss is reflected in a high probability of long-run loss. Suppose that the agent repeatedly violates C2 for θ_i ($i = 1, 2, \ldots$) that are probabilistically independent with respect to P^*. Thus, she sets $P_\mathcal{E}(\theta_i) = q_i \neq x_i = P^*(\theta_i)$. Assume further that the violations of C2 are non-negligible, that is, for all i, $|q_i - x_i| \geq \epsilon$ for some fixed $\epsilon > 0$. The agent's loss on each bet is $L_i = (q_i - I_{\theta_i})S_i$. Suppose the stake-maker sets $S_i = 1/(q_i - x_i)$. Then, the expected loss at each bet is $EL_i = (q_i - x_i)/(q_i - x_i) = 1$. Let $L^m = \sum_{i=1}^m L_i$ be the total loss after m bets, the expected total loss is $EL^m = \sum_{i=1}^m EL_i = m$. It is in the agent's interest that she will not lose out, that is, $L^m \leq 0$. Now, $P^*(L^m \leq 0) \leq P^*(|L^m - EL^m| > m)$. By the Chebyshev inequality, $P(|X - EX| > \alpha) \leq 1/\alpha^2 \mathit{Var} X$ for a random variable X and any probability function P (see, e.g. Billingsley 1979, pp. 80, 276). Hence, $P^*(L^m \leq 0) \leq 1/m^2 \mathit{Var} L^m$. Since the θ_i are independent, so are the L_i, and hence $\mathit{Var} L^m = \sum_{i=1}^m \mathit{Var} L_i$. But $\mathit{Var} L_i = (3x_i^2 + x_i)/(q_i - x_i)^2 \leq 4/\epsilon^2$. So $P^*(L^m \leq 0) \leq (1/m^2)(4m/\epsilon^2) = 4/m\epsilon^2 \longrightarrow 0$ as $m \longrightarrow \infty$. So the chance of avoiding long-run loss goes to zero. In sum, repeated non-negligible violations of C2 can be exploited to render the probability of long-run loss as close to 1 as you like. On the other hand, if the agent follows C2 then there are no violations to exploit and the expected loss is zero both in the short and the long run. So C2 should be followed to avoid long-run loss. Note that no assumption of uniform betting quotients is required for this justification.

To conclude, there are several ways one can argue for C2. One might take C2 to be a trivial defining characteristic of chance. Or, one might argue that chances are linked to frequencies and C2 is required to avoid sure loss in the long run (avoiding arbitrarily skewed degrees of belief). Or, in any case one can argue that C2 should be followed to avoid positive expected loss and to avoid extremely probable long-run loss.

3.3.1 Generalizing Calibration

Calibration as formulated by C2 deals with evidence of a single chance. In the remainder of this section we generalize the principle to cope with more complex evidence of chances and also to cope with non-quantitative structural evidence. Our aim will be to formulate a general Calibration norm of the form:

Calibration: $P_\mathcal{E} \in \mathbb{E}$,

In this case, the set of probability functions compatible with evidence, $\mathbb{E} \subseteq \mathbb{P}^\mathcal{L}$, is to be specified in an appropriate way.

We shall assume that evidence $\mathcal{E}$ narrows the chance function P^* down to a set $\mathbb{P}^* \subseteq \mathbb{P}^{\mathcal{L}}$ of probability functions. $\mathbb{P}^*$ should be thought of as defeasible: *current* evidence indicates that $P^* \in \mathbb{P}^*$ but future evidence may not. As to how cautious the agent should be about narrowing down $\mathbb{P}^*$ will depend on her current operating context. If a lot hangs on her being wrong about the chances, then $\mathbb{P}^*$ will be wide, that is, a large subset of $\mathbb{P}^{\mathcal{L}}$. Otherwise, it might be quite narrow: weak evidence that $P^*(\theta) = x$ may be sufficient to ensure that $\mathbb{P}^* \subseteq \{P \in \mathbb{P}^{\mathcal{L}} : P(\theta) = x\}$. Note that $\mathbb{P}^*$ may depend on various aspects of evidence $\mathcal{E}$. Clearly, it will depend on observations such as sample frequencies. But it will also depend on assumptions, for example, the assumption that the sample was selected from iid trials. Moreover, it will depend on theoretical considerations: in conjunction with theoretical considerations the observed symmetry of a coin may be enough to warrant granting that the coin has chance $\frac{1}{2}$ of landing heads; physical theory may be enough to alert the agent to the fact that chances of decay of a particle are unlikely to match a particular observed frequency in a small sample. The determination of $\mathbb{P}^*$, then, may be considered to be a part of a normative theory of evidence: given whatever else the agent grants, she ought to grant that $P^* \in \mathbb{P}^*$ for some suitable $\mathbb{P}^* \subseteq \mathbb{P}^{\mathcal{L}}$ that also depends on her current operating context as well as her language $\mathcal{L}$. Since the focus of this book is on the relation between evidence and belief, rather than on evidence itself, we will not dwell much on the determination of $\mathbb{P}^*$. It will suffice to note that frequentist statistics may be expected to play an important role in the determination of $\mathbb{P}^*$: frequentist statistics is concerned with determining one or more statistical models that are compatible with evidence $\mathcal{E}$; but such a set of statistical models can be construed as a subset $\mathbb{P}^*$ of $\mathbb{P}^{\mathcal{L}}$ within which the chance function is granted to lie. See §10.2 for further discussion of this point.

One might attempt to identify $\mathbb{E} = \mathbb{P}^*$, in which case the Calibration norm, $P_{\mathcal{E}} \in \mathbb{E}$, would become $P_{\mathcal{E}} \in \mathbb{P}^*$. But there are a few provisos which need taking into account. First,

C3: $\mathbb{E} \neq \emptyset$.

If evidence $\mathcal{E}$ is inconsistent, this tells us something about the evidence rather than about physical probability, so one cannot conclude that $\mathbb{P}^* = \emptyset$ and one can hardly insist that the agent have no beliefs at all, $P_{\mathcal{E}} \in \mathbb{E} = \emptyset$. Instead, $\mathbb{P}^*$ (and hence $\mathbb{E}$) must be determined by some consistency maintenance procedure. One might, to take a simple example of a consistency maintenance procedure, determine $\mathbb{P}^*$ from maximal consistent subsets of the evidence. More subtle consistency maintenance procedures can be employed to handle structured evidence: some evidence may be more entrenched than other evidence, in which case one would want, where possible, to rescind the latter before the former. Moreover, certain inconsistencies are most naturally handled by taking convex hulls. For example, if some evidence implies that $P^*(\theta) = 0.71$ but other, equally entrenched, evidence implies that $P^*(\theta) = 0.74$, then it is natural to consider both 0.71 and 0.74, as well as any value in between 0.71 and 0.74, as viable candidates for $P^*(\theta)$;

on the other hand, a value outside this range is unmotivated; hence, evidence arguably imposes the constraint $P^*(\theta) \in [0.71, 0.74]$. More generally, one can take $\mathbb{P}^*$ to be the convex hull of all chance reports. (The convex hull $\langle \mathbb{Q} \rangle$ of a set $\mathbb{Q}$ of probability functions is defined by $\langle \mathbb{Q} \rangle \stackrel{df}{=} \{\lambda Q + (1-\lambda)Q' : Q, Q' \in \mathbb{Q}, \lambda \in [0,1]\}$.) Whatever the consistency maintenance procedure, it is important to note that inconsistency of evidence does not imply that $\mathbb{P}^*$ is empty.

Inconsistency aside, there are apparently situations in which $\mathbb{P}^*$ is, in fact, empty. Consider the well-formed sentence θ: 'colourless green ideas sleep furiously'. It is a matter of debate as to whether the chance of this sentence being true is zero ($P^*(\theta) = 0$) or undetermined. If undetermined, it might be unconstrained ($P^*(\theta) \in [0,1]$) or undefined. If the latter, there is no chance function definable over a language in which this sentence can be formulated. We simply set aside any such example on the grounds that the sentence is meaningless: it does not express any proposition and, since propositions rather than sentences are the primary objects of both chance and belief, a Calibration norm of the form $P_\mathcal{E} \in \mathbb{P}^*$ remains viable at a semantic level if not at a syntactic level. One can salvage the syntactic level by simply stipulating that the chance of a sentence is unconstrained where the sentence expresses no proposition, and this is the approach we adopt here.

But there is a worry that this phenomenon may extend to meaningful sentences too. It has been argued that the frequency interpretation of probability only applies to cases in which events are repeated again and again (von Mises 1928, p. 11), and some notions of chance are similarly intended to apply only in cases where a single-case event is repeatable (Gillies 2000a, chapter 6). (This applies to theories of chance that are characterized in terms of a connection to frequency rather than to belief.) Hence, the present chance that in the next year there will begin a war that comes to be known as the 'third world war' might be considered to be undefined on the grounds that in the future it would be practically impossible to reconstruct present-day circumstances in order to generate a frequency to test the chance. In response to this worry, it suffices to accept that such chances may be untestable—and even unknowable—but point out that this does not imply that they are undefined. It is not too much of a stretch to agree that the chance of a third world war beginning in the coming year *varies* from year to year, even if one might disagree as to what the chance is. (In particular, the chance was close to zero before the second world war—it was then extremely unlikely that a second world war would begin and end and a further world war would begin, all within the space of a year.) Moreover, the present conditions for war, while not repeatable in practice, are repeatable in principle. Now, that some chances are unknowable is no problem for the Calibration norm, which merely requires that degrees of belief should accord with the chances that the agent grants.

While C3 is consistent with the identification $\mathbb{E} = \mathbb{P}^*$, there is a second proviso that casts doubt on this identification. Suppose that one grants that chance of θ is *either x or y*, where $y \neq x$. Then, $\mathbb{P}^* \subseteq \{P : P(\theta) = x \text{ or } P(\theta) = y\}$. But it seems ridiculous to insist that one should believe θ either to degree x or degree y. For example, take θ to posit the occurrence of some past event. Because it is an event in the past, the truth of θ is decided, and so θ has present chance either 0 or 1. But it is ridiculous to insist that one should either fully believe θ or fully disbelieve θ, unless, of course, one has some further evidence that forces this choice. One might suggest that one should simply disregard disjunctive evidence like this. But that will not do: suppose it is granted that the chance of θ is 0.71 or 0.74; disregarding such evidence would disregard the valuable information that θ is quite likely. If the convex hull approach to consistency maintenance is adopted, then it is natural to apply the convex hull approach here too:[4] if $\mathcal{E}$ implies $P, Q \in \mathbb{P}^*$, if $\mathbb{P}^*$ is most pertinent to $\mathcal{L}$, and if $R = \lambda P + (1 - \lambda)Q$ for $\lambda \in [0, 1]$, then $R \in \mathbb{E}$. Here, by saying that $\mathbb{P}^*$ is most pertinent to $\mathcal{L}$ we mean that if the agent were to grant that $P^* \in \mathbb{P}^*$, then that piece of evidence would be most pertinent to each θ on $\mathcal{L}$. Thus, we have

C4: If $\mathbb{P}^*$ is most pertinent to $\mathcal{L}$, then $\langle \mathbb{P}^* \rangle \subseteq \mathbb{E}$.

While C4 rules certain probability functions in as candidate belief functions, it does not rule any functions out. Hence, we need a further condition:

C5: $\mathbb{E} \subseteq \langle \mathbb{P}^* \rangle$.

Taking C4 and C5 together, as long as $\mathbb{P}^*$ is most pertinent to $\mathcal{L}$, $\mathbb{E} = \langle \mathbb{P}^* \rangle$. In which case, under the understanding that $\mathbb{P}^* \neq \emptyset$, C3 holds. C1 also holds as follows. If the agent grants θ and $\mathcal{E}$ is consistent, then $\mathbb{P}^* \subseteq \{P : P(\theta) = 1\}$. Now, by the definition of convexity it also follows that $\langle \mathbb{P}^* \rangle \subseteq \{P : P(\theta) = 1\}$.[5] Hence, $P_\mathcal{E}(\theta) = 1$. For similar reasons, C2 also holds.

---

[4]See Levi (1980, §9.5) for a rather different justification of convexity. Kyburg Jr and Pittarelli (1996, §4.A) express concerns about taking convex hulls in the presence of disjunctive constraints, arguing that long-run betting in such cases is likely to lead to loss. Clearly, if θ has chance 0 or 1, then betting at rate $\frac{1}{2}$ on θ over and over again may lead to loss (though not sure loss: as pointed out in §5.4 there is no scope for a Dutch book argument here). But that is no reason to preclude initially believing θ to degree $\frac{1}{2}$. First, that one is prepared to bet now with quotient $\frac{1}{2}$ on θ does not commit one to infinitely many bets at the same rate: presumably as results of previous bets are obtained, one's betting quotient will change. Second, betting quotient $\frac{1}{2}$, while opening oneself up to loss, may yet be a good quotient for controlling the extent of the loss. In §3.4, we see that such equivocal degrees of belief may be justifiable on the grounds that they minimize worst-case expected loss. Hence, convexity is especially plausible in the light of the Equivocation norm.

[5]A technical point. One can define a convex hull of a set of probability functions in various ways, depending on how the probability functions are themselves parameterized. Given the way the Probability norm was introduced in §3.2 it is natural (and quite standard) to define $R = \lambda P + (1 - \lambda)Q$ by $R(\omega) = \lambda P(\omega) + (1 - \lambda)Q(\omega)$. But one might define a probability function in other ways than by its values on the atomic states ω of the form $\pm A_1 \wedge \cdots \wedge \pm A_n$. For example, one might define a probability function P by the values it gives to $P(\pm A_i | \pm A_1 \wedge \cdots \wedge \pm A_{i-1})$, for $i = 1, \ldots, n$. In general, one can not expect different parameterizations to yield the same convex hulls (see, e.g. Haenni et al. 2010, §8.2.1). One of the advantages of the standard parameterization (which appeals to atomic states) is that it does yield C1: if P and Q give probability 1 to θ, then so will any convex combination of P and Q.

It remains to say when information about chances might not be most pertinent. It is important to distinguish between pertinence of evidence in $\mathcal{E}$ to a single proposition θ, considered earlier in formulating C2, and pertinence of $\mathbb{P}^*$ to $\mathcal{L}$, considered in the formulation of C4. In the former case, in a set of evidence of the form $\mathcal{E} = \{P^*(A_1 \to \neg A_2) \geq 0.9, A_1 \to \neg A_2\}$, the item of evidence $P^*(A_1 \to \neg A_2) \geq 0.9$ is *not* the evidence that is most pertinent to $A_1 \to \neg A_2$: since the agent also grants $A_1 \to \neg A_2$ itself, this item of evidence overrides the chance evidence. Taking $\mathcal{L}$ to be a finite propositional language on $A_1, A_2, \ldots, A_n$, then in this case $\mathbb{P}^* = \{P : P(A_1 \to \neg A_2) = 1\}$. Now $\mathbb{P}^*$ *is* most pertinent to $\mathcal{L}$, since there is no evidence in $\mathcal{E}$ that can be used to override or augment the claim that $P^* \in \{P : P(A_1 \to \neg A_2) = 1\}$.

But in some cases there can be pertinent evidence that does not impinge directly on chances. Consider $\mathcal{E} = \{P^*(A_3|A_1) = 0.9, P^*(A_3|A_2) = 0.8,$ both A_1 and A_2 are causes of A_3 but neither is a cause of the other$\}$. Plausibly, $\mathbb{P}^* = \{P : P(A_3|A_1) = 0.9, P(A_3|A_2) = 0.8\}$, that is, only the first two—quantitative—items of evidence are evidence of chances. Although the third—qualitative—item of evidence says nothing that can be directly transferred into a quantitative constraint on chances (we do not even know whether A_1 and A_2 are positive causes or preventatives of A_3), it does nevertheless constrain rational degrees of belief on $\mathcal{L} = \{A_1, A_2, A_3\}$. This is because of the following asymmetry of causality: while learning that two variables have a common cause can warrant a change in one's beliefs concerning those variables, learning that two variables have a common effect can not. For example, learning that two symptoms are in fact symptoms of the same disease can lead one to render those two symptoms more probabilistically dependent (with respect to one's rational belief function $P_\mathcal{E}$): one can reason that if a patient presents with the first symptom, she might well have the disease in question, in which case the second symptom is more plausible; hence, the first symptom becomes evidence for the second in the light of this new knowledge of the existence of a common cause. On the other hand, learning that two diseases turn out to have a common effect should not change one's degrees of belief concerning the two diseases: while one can reason that a patient who presents with one disease may well have the symptom in question, one cannot go on to conclude that that would make the second disease more credible, since the symptom would already be explained; hence, the first disease does not become evidence for the second in the light of new knowledge of a common effect. So qualitative causal evidence constrains degrees of belief in this way: if the agent learns of a new variable and learns that the variable is not a cause of any of the previous variables, then she should not change her degrees of belief concerning the previous variables (unless, of course, at the same time she also gains other evidence that *does* warrant such a change). Returning to our example, this implies that the evidence concerning A_3 (i.e. all three items of evidence in $\mathcal{E}$) should not change the agent's degrees of belief in A_1 and A_2 that would be obtained in the absence of such evidence: $P_\mathcal{E}^\mathcal{L}|_{\{A_1,A_2\}} = P_\emptyset^{\{A_1,A_2\}}$, that is, her belief function $P_\mathcal{E}^\mathcal{L}$, when restricted just to the variables A_1 and A_2, should match $P_\emptyset^{\{A_1,A_2\}}$, the belief function she would have had in the absence of A_3 and its evidence. Similarly, her causal evidence implies that evidence concerning A_2 has no bearing on A_1, $P_\emptyset^{\{A_1,A_2\}}\!\downharpoonright_{\{A_1\}} = P_\emptyset^{\{A_1\}}$, and vice versa, $P_\emptyset^{\{A_1,A_2\}}\!\downharpoonright_{\{A_2\}} = P_\emptyset^{\{A_2\}}$. In sum, the qualitative causal evidence in $\mathcal{E}$ imposes quantitative equality constraints on rational degrees of belief that are not mediated by evidence of chances. We call such constraints *structural constraints* to distinguish them from the *chance constraints* imposed via $\mathbb{P}^*$. In this case, then, $\mathbb{P}^*$ is not on its own most pertinent to $\mathcal{L}$.

Any relation, evidence of which constrains degrees of belief in this way, is called an *influence relation*. While causal influence is one example of an influence relation, other examples include logical influence, semantic influence, and various hierarchical relationships. Influence relations are explored in some detail in Williamson (2005a, §§5.8,10.5,11.4); in this section we just summarize the consequences of evidence of influence relationships. Suppose that it is reasonable to infer from evidence $\mathcal{E}$ that certain elementary propositions in $\mathcal{L}$ are *not* influences of the remaining elementary propositions $\mathcal{L}'$: we say that $\mathcal{L}'$ is *ancestral* with respect to $\mathcal{L}$ and $\mathcal{E}$. Let $\mathcal{E}'$ be the evidence in $\mathcal{E}$ that concerns only propositions in $\mathcal{L}'$. Let $\mathbb{P}^*$ be the set of potential chance functions as determined by $\mathcal{E}$ and $\mathbb{P}^{*\prime}$ be that determined by $\mathcal{E}'$. $\mathbb{P}^*$ is *compatible* with $\mathbb{P}^{*\prime}$ if for every probability function (on $\mathcal{L}'$) in $\mathbb{P}^{*\prime}$ there is a probability function (on $\mathcal{L}$) in $\mathbb{P}^*$ that extends it. We then have the following principle:

Irrelevance: If $\mathcal{L}'$ is ancestral with respect to $\mathcal{L}$ and $\mathcal{E}$ and $\mathbb{P}^*$ is compatible with $\mathbb{P}^{*\prime}$ then $\mathcal{L}\backslash\mathcal{L}'$ is *irrelevant* to $\mathcal{L}'$, that is, $P_\mathcal{E}^\mathcal{L}\!\downharpoonright_{\mathcal{L}'} = P_{\mathcal{E}'}^{\mathcal{L}'}$.

If $\mathcal{L}'$ is ancestral with respect to $\mathcal{L}$ and $\mathcal{E}$ and $\mathbb{P}^*$ is compatible with $\mathbb{P}^{*\prime}$ then $\mathcal{L}'$ is called a *relevance set* with respect to $\mathcal{L}$ and $\mathcal{E}$. Let $\mathcal{L}_1,\ldots,\mathcal{L}_k$ be all the relevance sets with respect to $\mathcal{L}$ and $\mathcal{E}$. Then, $\mathcal{E}$ imposes structural constraints $\mathbb{S}_\mathcal{E}^\mathcal{L} = \{P_\mathcal{E}^\mathcal{L}\!\downharpoonright_{\mathcal{L}_i} = P_{\mathcal{E}_i}^{\mathcal{L}_i} : i = 1,\ldots,k\}$. Now, there may well be relevance sets with respect to $\mathcal{L}_i$ and $\mathcal{E}_i$ too, for $i = 1,\ldots,k$, which in turn impose structural constraints $\mathbb{S}_{\mathcal{E}_i}^{\mathcal{L}_i}$. We take the set of structural constraints $\mathbb{S}$ to be the recursive closure of these constraints, that is, $\mathbb{S} = \mathbb{S}_\mathcal{E}^\mathcal{L} \cup \mathbb{S}_{\mathcal{E}_1}^{\mathcal{L}_1} \cup \cdots \cup \mathbb{S}_{\mathcal{E}_k}^{\mathcal{L}_k} \cup \cdots$. This gives us the final aspect of the Calibration principle:

C6: $\mathbb{E} \subseteq \mathbb{S}$.

If we suppose that structural and chance constraints exhaust the ways in which evidence constrains rational degree of belief, then C4–6 motivate the following explication of a Calibration norm:

Calibration: $P_\mathcal{E} \in \mathbb{E} = \langle \mathbb{P}^* \rangle \cap \mathbb{S}$.

Again C1–3 also hold under this conception of Calibration.

We make this supposition throughout the book, leaving the question of whether there are further ways in which evidence constrains rational belief as a question for future research.

3.3.2 Elicitation and introspection

We should note that the Calibration norm is often the hardest norm to apply in practice. It requires determining $\mathbb{E}$, which quantifies the impact of evidence on degrees of belief. To do this effectively, one needs to know, first, what one's evidence *is*, and second, what that evidence *implies* about chances and influences, in order to determine $\mathbb{P}^*$ and $\mathbb{S}$. In some circumstances, for example, devising AI systems or writing scientific papers, it is quite common to be explicit about what the evidence is and also to explore the consequences of evidence in some detail. But in other circumstances, for example, deciding whether or not to run for the train, it is hard to see how one might begin to calibrate. Clearly, in deciding whether or not to run for the train, time constraints may preclude any explicit decision-making process. But the question arises as to whether objective Bayesianism has anything to say in situations in which it is hard to be explicit about what one grants or hard to draw the consequences of what one grants.

Here, we can just reiterate some remarks made in §2.2: elicitation or introspection of degrees of belief is always available as a fallback option. Suppose that someone who you grant is rational has degree of belief in θ that is not entirely equivocal: say $P'(A_i) = x > P_=(\theta)$ where P' is that person's belief function and $P_=$ is the equivocator function defined by giving each atomic state ω the same probability. That the person is rational has two consequences in particular: he is right in granting what he does in fact grant, and his degrees of belief are appropriate given this evidence of his (§1.4). The latter consequence implies that he is a good objective Bayesian who adopts a maximally equivocal probability function P' in $\mathbb{E}'$ as his belief function: he has evidence that forces $P'(\theta) \geq x$, that is, $\mathbb{E}' \subseteq \{P : P(\theta) \geq x\}$. The former consequence implies that you should trust this evidence: hence your own $\mathbb{E}$ should be a subset of $\{P : P(\theta) \geq x\}$. So, even if you have no good evidence, or you are unsure what your evidence is or implies, you can narrow down $\mathbb{E}$ by eliciting probabilities from those you grant are rational. Further, if you have evidence that *your own* degrees of belief obtained by introspection are rational—for example, if, when you calculate what degrees of belief you ought to adopt, these tend to tally with your hunches—then introspection can similarly be used to narrow down $\mathbb{E}$ when applying the objective Bayesian method.

That the Calibration norm is hard to implement helps to explain why different individuals, apparently rational and apparently with the same evidence, might have different degrees of belief. As Laplace noted, probability depends not just on data but also on knowledge of the implications of data:

> The difference of opinions depends, however, upon the manner in which the influence of known data is determined. The theory of probabilities holds to considerations so delicate that it is not surprising that with the same data two persons arrive at different results. (Laplace 1814, p. 10)

Jakob Bernoulli, too, recognized that measuring probabilities is something of an art:

> To *conjecture* about something is to measure its probability. Therefore we define the *art of conjecture*, or *stochastics*, as the art of measuring the probabilities of things as

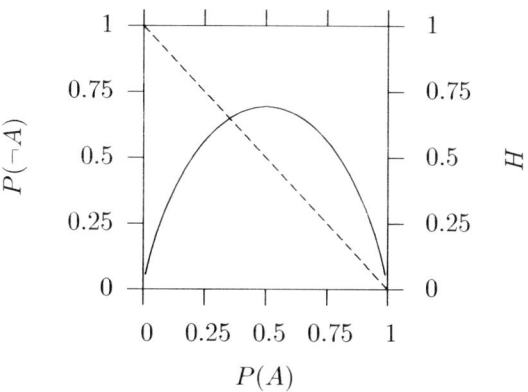

FIG. 3.1. Probability functions (dotted line) and their entropy H (solid curve) in two dimensions.

exactly as possible, to the end that, in our judgements and actions, we may always choose or follow that which has found to be better, more satisfactory, safer, or more carefully considered. On this alone turns all the wisdom of the philosopher and all the practical judgement of the statesman. (Bernoulli 1713, p. 213)

In sum, applying the Calibration norm can require some subtlety, but that is no argument against the norm itself.

3.4 Equivocation

The Equivocation norm says that degrees of belief should, apart from being probabilities and satisfying constraints imposed by evidence, be equivocal. That is, where evidence fails to fully determine appropriate degrees of belief, the agent should adopt a belief function sufficiently close to the equivocator:

Equivocation: $P_\mathcal{E} \in \Downarrow\mathbb{E}$.

As discussed in §2.3, $\Downarrow\mathbb{E}$ is the pragmatically determined set of probability functions that are sufficiently close to $P_=$. The equivocator is defined by $P_=(\pm A_1 \wedge \cdots \wedge \pm A_n) = 1/2^n$. Here, closeness is understood in terms of divergence $d(P,Q) = \sum_\omega P(\omega) \log P(\omega)/Q(\omega)$. Letting $\downarrow\mathbb{E} \stackrel{\text{df}}{=} \{P \in \mathbb{E} : d(P, P_=) \text{ is minimized}\}$, we have that $\downarrow\mathbb{E} = \{P \in \mathbb{E} : \text{the entropy } H(P) = -\sum_\omega P(\omega) \log P(\omega) \text{ is maximized}\}$. The set $\Downarrow\mathbb{E}$ of sufficiently equivocal evidentially compatible probability functions should contain the set $\downarrow\mathbb{E}$ of maximally equivocal functions, $\downarrow\mathbb{E} \subseteq \Downarrow\mathbb{E}$. But $\Downarrow\mathbb{E}$ may in general contain other functions too, and it certainly will in cases in which there is no maximally equivocal function. Note that since $\mathbb{E}$ is convex, if $\downarrow\mathbb{E}$ is non-empty, then it is a singleton.

If there is a single elementary proposition $\mathcal{L} = \{A\}$, then there are two atomic states $\Omega = \{A, \neg A\}$ and the Probability norm requires that $P(\neg A) = 1 - P(A)$, as depicted by the dotted line in Fig. 3.1. Entropy H is depicted by the solid

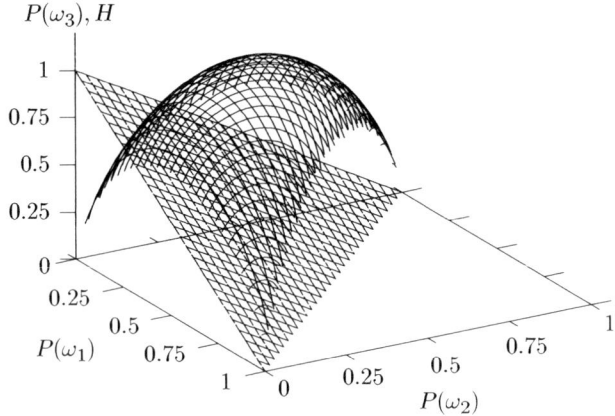

FIG. 3.2. Probability functions (plane) and their entropy H (curved surface) in three dimensions.

line. Clearly, the closer to the centre of the dotted line, the higher the entropy. Strictly speaking, there is no three-dimensional analogue of this picture since there are always an even number of atomic states, but to help engender some higher-dimensional intuition, Fig. 3.2 would be the corresponding visualization of a three outcome case, $\Omega = \{\omega_1, \omega_2, \omega_3\}$. The probability functions are depicted by the plane and their entropy by the curved surface.

In this section, we look at the motivation for going beyond empirically based subjectivism (distinguished by its appeal to Probability and Calibration) to objective Bayesianism, which adopts Equivocation too.

Epistemologists who focus primarily on qualitative beliefs rather than degrees of belief (*the agent believes θ* rather than *the agent believes θ to degree x*) typically take the qualitative analogue of the Equivocation norm to be uncontroversially true. In qualitative terms, Equivocation says that an agent should believe θ if and only if her evidence forces a belief in θ. The qualitative analogue of the subjectivist position, on the other hand, would hold that an agent is rational to believe θ if θ is consistent with her evidence. Yet the latter kind of permissive rationality is barely considered by qualitative-belief epistemologists. Thus, for example, 'ordinarily it is rational for you, all things considered, to believe only those hypotheses for which you have adequate evidence' (Foley 1993, p. 22). Surprisingly, perhaps, the boot is on the other foot when it comes to quantitative beliefs: subjectivism—taken to include strict subjectivism and empirically based subjectivism, both of which reject Equivocation—is far more widely advocated by quantitative epistemologists than objective Bayesianism, which endorses Equivocation as well as the Probability and Calibration norms. While many share the intuition behind the Equivocation norm and agree that there must be something right about it, it is notoriously hard to put one's finger on exactly what is right about it. This confusion about motivation—taken together with the perceived problems for the Maximum Entropy Principle, addressed elsewhere in this book—helps to explain the surprising conflict between qualitative and quantitative epistemology. Here, we consider a range of ways in which one might motivate the Equivocation norm and argue that one of these putative justifications does indeed point to what is right about the norm. With the prospect of coherent motivation, the Equivocation norm can be taken seriously—the qualitative epistemologist's less permissive take on rationality can be extended to the quantitative case.

First, we shall consider the two standard arguments in favour of the Maximum Entropy Principle—an argument by interpretation and an argument by derivation. As we see, neither of these arguments conclusively decides between objective Bayesianism and empirically based subjectivism.

The original justification of the Maximum Entropy Principle in Jaynes (1957a) is perhaps best known. This justification appeals to Claude Shannon's use of entropy as an explication of the uncertainty embodied in a probability function (Shannon 1948, §6). Jaynes maintains that an agent's belief function should be informed by evidence but should otherwise be maximally uncertain or non-committal—thus it should have maximum entropy according to Shannon's measure:

> in making inferences on the basis of partial information we must use that probability distribution which has maximum entropy subject to whatever is known. This is the only unbiased assignment we can make; to use any other would amount to arbitrary assumption of information which by hypothesis we do not have.
> ...
> The maximum entropy distribution may be asserted for the positive reason that it is uniquely determined as the one that is maximally noncommittal with regard to missing information (Jaynes 1957a, p. 623).

While the interpretation of uncertainty via entropy is plausible and now routinely adopted throughout the sciences, there is a gap in the argument. Jaynes *assumes* that maximally non-committal, unbiased degrees of belief are most desirable and argues that they should then be found by maximizing entropy. Even if we grant Jaynes that entropy measures lack of commitment, we still need some reason to accept his premise. Why should maximally non-committal degrees of belief be any better than, say, maximally committal degrees of belief? Why is bias bad?

The second key line of argument in favour of the Maximum Entropy Principle takes the form of a derivation. There are various versions, but the derivation of Jeff Paris and Alena Vencovská is perhaps most compelling (Paris and Vencovská 1990, 2001; Paris 1994). Their argument takes the following form:

- An *inference process* is a function which maps a language and evidence to a probability function on the language that satisfies the evidence.
- If the selected probability function is to be construed as a representation of the degrees of belief that one ought to adopt on the basis of the evidence, then the inference process ought to satisfy some commonsense desiderata. For example, given two logically equivalent evidence bases the inference process should select the same probability function.
- The only inference process that satisfies these desiderata is the Maximum Entropy Principle.
- Therefore, the only commonsensical inference process is the Maximum Entropy Principle.

Again, even if we grant that the only inference process satisfying the desiderata is the Maximum Entropy Principle, this argument does not do enough for our purposes.

One worry—which can, to some extent, be allayed—is that the argument assumes from the outset that we need an inference process, that is, that we need to select a single probability function that satisfies the evidence and that this probability function must be uniquely determined by the language and evidence. Subjectivists would disagree with both these assumptions, arguing that different individuals are free to choose different belief functions on the basis of the same evidence. This worry can be allayed because Paris and Vencovská (2001, §3) in fact relax the uniqueness requirement when they consider the case in which background knowledge imposes non-linear constraints on degrees of belief. They do this because non-linear constraints may describe a non-convex set, in which there may be no unique entropy maximizer (Paris and Vencovská do not take the convex hull of the set described by constraints as we do here). What they show is that even without an assumption of uniqueness of belief function, the desiderata still isolate the Maximum Entropy Principle as the only commonsensical inference process.

But whether or not the uniqueness requirement is relaxed, the argument is unlikely to satisfy the subjectivist. This is because it is an argument by derivation: strong assumptions are required to derive a strong conclusion, and one can easily deny the latter by denying the former. The Renaming Principle, for instance, dictates that (evidence permitting) degrees of belief should be invariant under permutations of the language; the Independence Principle holds that in certain cases degrees of belief should be independent if there is no evidence of dependence. While these desiderata may be fruitful, they are far from trivial, and more justification is required to convince the proponent of empirically based subjectivism.[6]

---

[6]Hosni and Paris (2005) explore a line of justification of the desiderata, claiming that they are commonsensical because they force us to assign similar probabilities, and that conformity is some kind of rational norm. Again, many would take issue with these claims. Assigning similar probabilities may be of pragmatic advantage, but hardly seems to be a requirement of

So we see that current justifications of the Maximum Entropy Principle do not fully motivate the move to objective Bayesianism. The standard argument by interpretation presumes that an agent's degrees of belief should be maximally non-committal. On the other hand, the standard argument by derivation is an argument by derivation, and hence unlikely to convince a detractor. Further argument is needed.

3.4.1 The argument from objectivity

One might try to construct an argument around the suggestion that empirically based subjective Bayesianism is *not objective enough* for many applications of probability.[7] If applications of probability require full objectivity—that is, a single probability function that fits available evidence—then perhaps one can argue that the maximum entropy function is the only appropriate choice.

Subjective Bayesian method typically involves the following process, called *Bayesian conditionalization* (see also Chapter 4). First, a *prior* probability function P must be produced. Then, new evidence ε is collected. Finally, predictions are drawn using the *posterior* probability function $P'(\theta) = P(\theta|\varepsilon)$. Now, the prior function is determined before the new evidence is available; this is entirely a matter of subjective choice for empirically based subjectivists. However, the ensuing conclusions and predictions will typically be sensitive to this initial choice of prior, rendering them subjective too. If, for example, $P(\theta) = 0$, then $P'(\theta) = P(\theta|\varepsilon) = P(\varepsilon \wedge \theta)/P(\varepsilon) = 0$ (assuming $P(\varepsilon) > 0$). On the other hand, if $P(\theta) = 1$, then $P'(\theta) = P(\varepsilon|\theta)P(\theta)/(P(\varepsilon|\theta)P(\theta) + P(\varepsilon \wedge \neg\theta)) = P(\varepsilon|\theta)/P(\varepsilon|\theta) = 1$.

It is plain to see that two subjectivists can radically disagree as to the conclusions they draw from evidence. If they adopt the opposite prior beliefs, they will draw the opposite conclusions. Under the subjectivist (strictly subjectivist or empirically based subjectivist) approach, if conditionalization is adopted, then all conclusions are relative to initial opinion.

If the empirically based subjectivist is to avoid such strong relativity, she must reject Bayesian conditionalization as a universal rule of updating degrees of belief. The standard alternative is *cross-entropy updating* (Williams 1980; Paris 1994, pp. 118–26). Here the agent adopts prior belief function P and her posterior P' is taken to be the probability function, out of all those that are compatible with the new evidence, that is closest to P in terms of the divergence function d defined above. (Note that, unlike the Bayesian conditionalization case, the evidence does not have to be representable as a sentence ε of the agent's language.)

Suppose $\Omega = \{A, \neg A\}$, for instance. Two empirically based subjectivists may set $P(A) = 0$ and $Q(A) = 1$, respectively, while an objective Bayesian is forced to

---

rationality except in special cases (Gillies 1991). Nor need this consequence of the desiderata render the desiderata commonsensical in themselves—the end does not justify the means in this case.

[7] Note that Salmon (1990, §4) argues for a version of empirically based subjective Bayesianism on the grounds that subjective Bayesianism is not objective enough for science.

set $R(A) = 1/2$ in the absence of any empirical evidence. Now, suppose evidence is collected that constrains the probability of A to lie in the interval $[0.6, 0.9]$. The objective Bayesian must adopt $R'(A) = 0.6$ while the empirically based subjectivists now adopt posteriors $P'(A) = 0.6$ and $Q'(A) = 0.9$, respectively. Thus, with cross-entropy updating, evidence can shift degrees of belief away from 0 and 1.

While cross-entropy updating may be an improvement over Bayesian conditionalization, one might argue that the remaining relativity in empirically based subjective Bayesianism is still too much for applications of probability if applications demand full objectivity.

But such an argument would be hard to execute. First, one would expect that the amount of tolerable subjectivity would vary from application to application—it is unlikely to be the case that *all* applications of probability demand full objectivity. While objectivity of conclusions seems desirable in a computer system for controlling nuclear retaliation, it is clearly less desirable in an automated oenophile.

Second, it is difficult to judge the need for objectivity. Scientists often emphasize the objectivity of their disciplines, but it can be difficult to say whether their claims reflect their science's need for objectivity or their own. (Of course, a perceived objectivity of science is rhetorically very useful to scientists—conclusions appear more forceful.) Moreover, even if scientific methodology does assume an inherent objectivity, such an assumption may simply be erroneous. There may be less objectivity to be found than commonly supposed (Press and Tanur 2001). These difficulties have led to widespread disagreement between sociologists of science on the one hand, many of whom view scientific conclusions as highly relative, and philosophers of science and scientists on the other hand, many of whom view science as an objective activity by and large (Gottfried and Wilson 1997). Until some (objective) common ground can be found in the study of science, we are a long way from determining whether the extreme objectivity of objective Bayesianism is required in science, or whether the more limited objectivity of empirically based subjective Bayesianism is adequate.

Finally, even in cases where objectivity is required, that objectivity need not necessarily incline one towards objective Bayesianism. In order to run the axiomatic derivation of the Maximum Entropy Principle, one must first accept that the commonsense desiderata are indeed desirable. If not, one may be led to alternative implementations of the objectivity requirement. Consider the *Minimum Entropy Principle*: if $\mathcal{L} = \{A_1, \ldots, A_n\}$ choose as belief function the probability function, out of all those that satisfy constraints imposed by evidence $\mathcal{E}$, that is as close as possible to the function P that sets $P(A_1) = 1, P(A_2) = \cdots = P(A_n) = 0$. This is objective in the sense that once the language and its ordering have been fixed, then so too is the belief function. Moreover, this principle is not as daft as it may first look—there may be semantic reasons for adopting such an unbalanced prior. Consider a criminal trial setting: $\mathcal{L} = \{A_1, A_2\}$, where A_1 represents innocence and A_2 represents guilt, and where no evidence has been presented as yet—then the Minimum Entropy Principle represents a prior assumption of innocence; this appears to be the recommended prior.

In sum, while it may be tempting to argue for objective Bayesianism on the grounds that applications of probability demand objective conclusions, there are several hurdles to be overcome before a credible case can be developed.

3.4.2 Rationality and evidence

One thing should be clear by now: it can not be the empirical warrant that motivates the selection of a particular belief function from all those compatible with evidence, since all such belief functions are equally warranted by available evidence. If objective Bayesianism is to be preferred over subjectivist approaches, it must be for non-evidential reasons. (Equally, one cannot cite evidence as a reason for abandoning objective Bayesianism in favour of a subjectivist approach.)

Thus, the problem of deciding between objective Bayesianism and empirically based subjectivism hinges on the question of whether evidence exhausts rationality. Objective Bayesianism supposes that there is more to rationality than evidence: a rational agent's degrees of belief should not only be well calibrated with evidence, but they should also be sufficiently middling. For empirically based subjectivism, on the other hand, empirical warrant is sufficient for rationality. (This puts the empirically based subjectivist at a disadvantage, because from the empirical perspective there are simply no considerations that can be put forward to support empirically based subjectivism over objective Bayesianism since both are empirically optimal; in contrast, the objective Bayesian can proffer non-empirical reasons to prefer objective Bayesianism over empirically based subjectivism.)

If rationality goes beyond evidence, what else does it involve? We have already discussed one form of non-evidential reason that might decide between the two types of approach—a demand for *objectivity*. But there are other more overtly pragmatic reasons that can be invoked to the same end. Shortly we shall see whether *caution* can be used to motivate objective Bayesianism. But first we turn to *efficiency*.

3.4.3 The argument from efficiency

One might be tempted to appeal to efficiency considerations to distinguish between objective Bayesianism and empirically based subjective Bayesianism. If objective Bayesian methods are more efficient than empirically based subjective Bayesian methods then that would provide a reason to prefer the former over the latter.

One possible line of argument proceeds as follows. If probabilities are subjective, then their measurement requires finding out which particular degrees of belief have been chosen by some agent. This can only be done by elicitation: asking the agent what her degrees of belief are, or perhaps inducing them from her behaviour (e.g. her betting behaviour). But, as developers of expert systems in AI have found, elicitation and the associated consistency-checking are prohibitively time-consuming tasks (the inability of elicitation to keep pace with the demand for expert systems was known as *Feigenbaum's bottleneck*). If subjective Bayesianism is to be routinely applied throughout the sciences, it is likely that a similar bottleneck will be reached. For example, determining the most probable statistical model given evidence would first require eliciting model assumptions (Are the agent's degrees of belief normally distributed, for instance? Are certain degrees of belief probabilistically independent?) and also the agent's prior degree of belief in each model—a daunting task. On the other hand, if objective Bayesianism is adopted, degrees of belief are objectively determined by evidence and elicitation is not normally required—degrees of belief are calculated by maximizing entropy. Therefore, objective Bayesianism is to be preferred for reasons of efficiency.

Of course, this argument fails if the objective Bayesian method is itself more computationally intractable than elicitation. Indeed, Judea Pearl rejected the Maximum Entropy Principle on the grounds that computational techniques for maximizing entropy were usually intractable (Pearl 1988, p. 463). However, while that was indeed the case in 1988, it is not the case now. Pearl's own favoured computational tool, *Bayesian nets*, can be employed to vastly reduce the complexity of entropy maximization, rendering the process tractable in a wide variety of natural settings—see Chapter 6. Thus, despite Pearl's doubts, efficiency considerations do lend support to objective Bayesianism after all.

But efficiency considerations on their own fail to distinguish between objective Bayesianism and other procedures for selecting a unique probability function. The Maximum Entropy Principle is no more efficient than the Minimum Entropy Principle. Worse, consider the *Blind Minimum Entropy Principle*, where one ignores evidence and minimizes entropy straight off: if $\mathcal{L} = \{A_1, \ldots, A_n\}$ choose as belief function the probability function P such that $P(A_1) = 1$ and $P(A_2) = \cdots = P(A_n) = 0$. This modified principle avoids elicitation and is far easier to implement than the Maximum Entropy Principle. Should one really blindly minimize entropy?

If not, efficiency cannot be the whole story. At best, one can say something like this: efficiency considerations tell against elicitation and motivate some procedure for mechanically determining an agent's degrees of belief; other desiderata need to be invoked to determine exactly which procedure; an axiomatic derivation might then be used to show that the Maximum Entropy Principle is the only viable procedure.

While this is an improvement over the argument from objectivity, it is still rather inconclusive as it stands. Further work needs to be done to explain why efficiency is not the whole story, that is, to explain why the other desiderata override efficiency considerations. The other desiderata thus play an important role in this new argument, and stand in need of some form of justification themselves. Given that subjectivists would find several of these desiderata dubious, their justification may turn out to be more of an ordeal than elicitation.

3.4.4 The argument from caution

As Ramsey notes, our degrees of belief guide our actions and our actions are tantamount to bets. To embark on a course of action (such as running to the station), a degree of belief (in the train running, in this case) must be sufficiently high. Now, every course of action has its associated risks. Running to the station only to find that the train is not running wastes time and effort, and one may miss an important appointment. Such a course of action is not to be embarked upon lightly, and prudence is required. The trigger for action will vary according to risk—if a lot hangs on it, one may only run to the station if one has degree of belief at least 0.95 in the train running, but if the consequences are less dire, a lower trigger-level, say 0.85, may be appropriate. Suppose one knows only that the local train-operating company has passed the minimum threshold of 80% of trains running. According to empirically based subjective Bayesianism, one's degree of belief in the train running can be chosen from anywhere in the interval $[0.8, 1]$. According to objective Bayesianism, the least extreme value in this interval must be chosen: 0.8. So the empirically based subjectivist may decide to go to the station while the objectivist decides not to. One can argue, then, that the objective Bayesian decision is more cautious and is to be preferred, since there is no empirical evidence to support a less cautious decision.

In sum, extreme degrees of belief trigger actions and open one up to their associated risks. In the train example, the objective Bayesian strategy of adopting the least extreme degree of belief seems to be the most prudent. Can one abstract from this particular case to argue that objective Bayesianism is to be preferred in general? There are some potential difficulties with such a move to generality, as we now see.

The first potential problem stems from the fact that it is not only extreme degrees of belief that trigger actions—middling degrees of belief can also trigger actions. Consider a case where a patient has one of the two possible diseases A and B. A high degree of belief in A will trigger a course of medication to treat A. Similarly, a high degree of belief in B will trigger treatment of B. However, a more middling degree of belief—a degree of belief that triggers neither treatment of A nor treatment of B—will trigger another action, perhaps the gathering of more evidence in order to reach a conclusive diagnosis. Collecting further symptoms from the patient also has its associated costs and risks: it requires time, effort, and money to perform more tests, and some tests might harm the patient. Now, objective Bayesianism advocates setting middling degrees of belief, thereby exposing the diagnoser to the risks associated with collecting more symptoms. It seems that objective Bayesianism is not such a risk-averse position after all.

But this apparent problem does not in fact scupper the prospect of a general argument in favour of objective Bayesianism. Granted, middling degrees of belief do often trigger actions—indeed always, since inactivity itself exposes an agent to associated risks and so can be considered an action in these sense intended here. But it may yet be the case that in real decision problems the risks associated with actions triggered by middling degrees of belief tend to be less than those triggered by extreme degrees of belief. Suppose in the above example that A and B are two types of meningitis, requiring very different treatment. The risks associated with either outcome are so high that the risks associated with collecting more symptoms pale into insignificance in comparison. Suppose on the other hand that A and B are just two strains of cold; A responds best to a nightly dose of rum toddy while B requires a hot toddy taken three times a day; in either case a full recovery will be made even if no treatment is taken. In this case, the risks associated with either diagnosis are so low that in the absence of a diagnosis it simply is not worth doing the blood test that would provide conclusive evidence: a more middling degree of belief triggers only inactivity. The point is that collecting further evidence will only be an option if the resulting evidence is worth it, that is, if the costs and risks associated with the primary outcomes A and B outweigh those associated with collecting new evidence. Hence, it will still be most cautious to adopt more middling degrees of belief.

It might be thought that there is a more serious variant of this problem. Here is the argument: for a politician the risks associated with appearing non-committal outweigh those of committing to an unjustified course of action or making a promise that cannot be kept; people like their politicians bold and it would seem that a non-committal objective Bayesian politician would not get elected. But this objection can be met. Even if it is true that politicians who appear non-committal are more likely to get elected, the objection hinges on a mistaken conflation of appearance and reality. The fact that a politician should not *appear* non-committal does not mean that her beliefs should not actually *be* non-committal—politicians simply need to mask their beliefs. Their beliefs need to be as cautious as anyone else's though: they need to be shrewd judges of which promises the electorate will swallow, and should not commit to one lie over another unless they have a justified belief that they can get away with it.

A second type of problem confronts any attempt to generalize the argument from caution. This concerns cases in which risks are known to be imbalanced. In the diagnosis example, take A to be meningococcal meningitis and B to be 'flu. In this case, the risks associated with failing to diagnose A when it is present are so great that it may be prudent to *assume* A and prescribe antibiotics, unless there is conclusive evidence that decides in favour of B. We have already come across another example of an imbalance of risks: the risk of judging the innocent guilty is considered to outweigh that of judging the guilty innocent, and this motivates a presumption of innocence in criminal cases. Such presumptions seem far from non-committal, yet rational.

Perhaps, the best way of resolving this difficulty is to distinguish appearance and reality again. It is important in these cases to *act as if* one believed one of the alternatives—to prescribe antibiotics or to release the suspect—not to actually believe that alternative. In these cases, the imbalanced risks motivate imbalanced trigger levels rather than imbalanced degrees of belief. If the degree of belief in 'flu is higher than, say 0.95, prescribe paracetamol, otherwise, if the degree of belief in meningococcal meningitis is at least 0.05, prescribe antibiotics. If guilt is not proved beyond reasonable doubt (if the degree of belief in guilt is less than, 0.99 say), then trigger action that corresponds to innocence, that is, release the suspect. In both these cases the trigger level for one alternative is very high while the trigger level for the other alternative is very low.

One might respond to this move by accepting this proposed resolution for the diagnosis example, but rejecting it for the legal example. One might claim that in the legal case there should not only be a high standard of reasonable doubt for guilt and a corresponding low standard of doubt for innocence, but there should also be prior degree of belief 1 in innocence, in order to make it as hard as possible for the prosecution to sway degree of belief to beyond a reasonable doubt. This response seems natural enough—it seems right to make it as hard as possible to trigger guilt. However, this response should not be acceptable to Bayesians who are sceptical about the Equivocation norm—whether strictly subjective or empirically based Bayesians—because it does not sit well with subjective Bayesian methods of updating. Recall that we saw that there are two standard subjectivist options for updating an agent's degree of belief in new evidence, Bayesian conditionalization, and cross-entropy updating. If Bayesian conditionalization is adopted, then a prior degree of belief 0 of guilt can never be raised above 0 by evidence, and it will be impossible to convict anybody. On the other hand, if cross-entropy updating is adopted, then a presumption of innocence will make no difference in the legal example. A presumption of innocence corresponds to prior degree of belief 0 in guilt, while a maximally non-committal probability function will yield a prior degree of belief of $\frac{1}{2}$. In either case, degree of belief can only be raised above 0.99 if empirical evidence constrains degree of belief to lie in some subset of the interval $[0.99, 1]$ (because the cross-entropy update is the degree of belief, from all those that are compatible with evidence, that is closest to 0). As long as the prior degree of belief is lower than the trigger level for guilt, triggering is dependent only on the evidence, not on prior degree of belief. In sum, whichever method of updating the subjectivist adopts, there is no good reason for setting prior degree of belief in guilt to be 0.

There is a third—more substantial—problem that besets an attempt to generalize the argument from caution: when we move to higher dimensions it becomes clear that the Maximum Entropy Principle is not *always* the most cautious policy for setting degrees of belief (see, e.g. Kyburg Jr and Pittarelli 1996, §2.E). To help visualize the problem, consider a case in which the agent's language $\mathcal{L} = \{A, B\}$, so there are four atomic states, $\omega_1 = \neg A \wedge \neg B, \omega_2 = A \wedge \neg B, \omega_3 = \neg A \wedge B$, and $\omega_4 = A \wedge B$. While on a two-dimensional space the set of all probability functions is a line segment (Fig. 3.1) and on a three-dimensional space the set of probability functions is a triangular segment of the plane (Fig. 3.2), on this four-dimensional space the set of all probability functions can be represented by a tetrahedron whose vertices represent the functions defined by $P(\omega_1) = 1, P(\omega_2) = 1, P(\omega_3) = 1$, and $P(\omega_4) = 1$, respectively, as depicted in Fig. 3.3. Suppose that a risky action is triggered if $P(A) \geq 5/6$

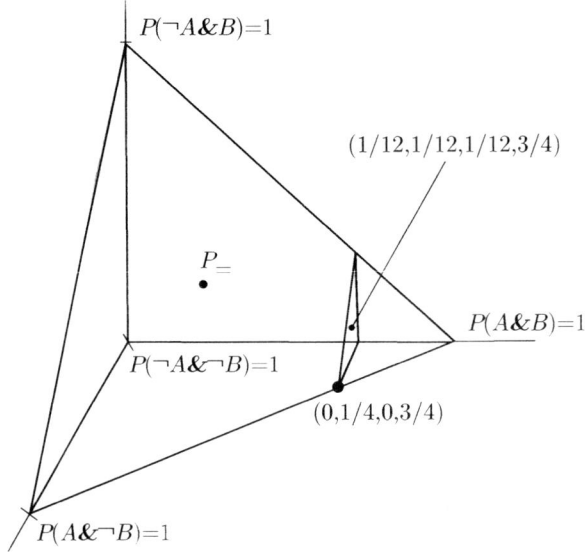

FIG. 3.3. Maximum and Minimum Entropy Principle probability functions, $(1/12, 1/12, 1/12, 3/4)$ and $(0, 1/4, 0, 3/4)$, respectively.

and another risky action is triggered if $P(B) \geq 5/6$. Suppose the evidence $\mathcal{E} = \{P^*(A \wedge B) \geq 3/4\}$. Then, the set $\mathbb{E}$ of probability functions that are compatible with this knowledge is represented by the smaller tetrahedron in Fig. 3.3. Now the Maximum Entropy Principle advocates adopting the probability function P from $\mathbb{E}$ that is closest to the equivocator $P_=$ which sets $P_=(\omega_1) = P_=(\omega_2) = P_=(\omega_3) = P_=(\omega_4) = 1/4$. Thus, the maximum entropy function sets $P(\omega_1) = P(\omega_2) = P(\omega_3) = 1/12$, and $P(\omega_4) = 3/4$. This triggers both A and B since $P(A) = 5/6$ and $P(B) = 5/6$. On the other hand, the Minimum Entropy Principle advocates adopting the probability function Q from $\mathbb{E}$ that is closest to the probability function $Q_<$ which sets $Q_<(\omega_2) = 1$. Thus the minimum entropy function sets $Q(\omega_1) = 0, Q(\omega_2) = 1/4, Q(\omega_3) = 0$, and $Q(\omega_4) = 3/4$.[8] This triggers A, since $Q(A) = 1 \geq 7/8$, but does not trigger B, since $Q(B) = 3/4 < 5/6$. Hence, in this case the Minimum Entropy Principle licences the most cautious course of action while the Maximum Entropy Principle seems to throw caution to the wind.

While this example shows that the most equivocal probability function is not the most cautious *in every situation*, it may yet be the most cautious *on average*. If it is most cautious when averaging over evidence bases $\mathcal{E}$ and decisions (i.e. sentences θ and trigger levels $\tau_\theta \in [0, 1]$ such that a course of action is triggered

---

[8]The 'Minimum Entropy Principle' is a bit of a misnomer—while the probability function Q commits most to ω_1, it does not actually have minimum entropy in this example. Here, entropy is in fact minimized by the probability function R that commits most to $A \wedge B$, $R(\omega_4) = 1$.

Motivation

if $P(\theta) \geq \tau_\theta$), then adopting the Maximum Entropy Principle will be the best policy in the absence of any prior knowledge about $\mathcal{E}, \theta$, and τ_θ. (We can leave it open here as to the mechanism that is used to select the trigger levels. These may be set by experts perhaps, by maximizing expected utility, or by some other decision-theoretic procedure.)

The average caution for a policy for setting belief function P can be measured by the proportion of $\mathbb{E}, \theta, \tau_\theta$ that result in a course of action being triggered. Define the *trigger function* $T(P, \theta, \tau_\theta) = 1 \Leftrightarrow P(\theta) \geq \tau_\theta$, and 0 otherwise. Then, the average caution of a policy is the measure of $\mathbb{E}, \theta, \tau_\theta$ for which $T(P, \theta, \tau_\theta) = 1$, for P determined by $\mathbb{E}$. Restricting attention to non-logically equivalent sentences, one can measure the average proportion of triggers for a policy that selects $P_\mathbb{E}$ from $\mathbb{E}$ by

$$|T| \stackrel{\text{df}}{=} \frac{1}{z} \sum_{\Omega_i \subseteq \Omega} \int_\mathbb{E} \int_{\tau_{\theta_i}} T(P_\mathbb{E}, \theta_i, \tau_{\theta_i}) d\mathbb{E} \, d\tau_{\theta_i}$$

where z is a normalizing constant, $z = 2^{|\Omega|} \int_\mathbb{E} \int_{\tau_{\theta_i}} 1 d\mathbb{E} \, d\tau_{\theta_i}$, and θ_i is defined as the disjunction of atomic states in $\Omega_i \subseteq \Omega$, with $\bigvee \emptyset$ taken to be a contradiction, say $A_1 \wedge \neg A_1$. The smaller the value of $|T|$, the more rarely actions are triggered and the more cautious the policy is on average.

However it turns out that the Maximum Entropy Principle is no more cautious than the Minimum Entropy Principle, even if one averages out using this measure. By way of example consider the two-dimensional case, $\Omega = \{A, \neg A\}$. Suppose $\mathcal{E}$ constrains $P(A)$ to lie in a closed interval $[x, y]$, where $0 \leq x \leq y \leq 1$, and that the trigger levels are all the same, $\tau_\theta = \tau \in [0, 1]$. We are interested in the proportion of values of x, y, τ that trigger a decision, that is, the volume of the part of the cube $[0, 1]^3$ defined by these parameters that triggers some decision. We are only concerned with the half of the cube such that $y \geq x$, so $z = 4 \times 1/2 \times 1 = 2$. Let the Ω_i run through the subsets of Ω: $\Omega_1 = \emptyset, \Omega_2 = \{A\}, \Omega_3 = \{\neg A\}$, and $\Omega_4 = \{A, \neg A\}$. Trivially, $\theta_4 = A \vee \neg A$ always gets triggered and $\theta_1 = A \wedge \neg A$ only gets triggered if $\tau = 0$. Consider first the entropy maximization policy. In this case $T(P_\mathbb{E}, A, \tau) = 1 \Leftrightarrow x \geq \tau$ or $\tau \leq 1/2$ and $y \geq \tau$, while $T(P_\mathbb{E}, \neg A, \tau) = 1 \Leftrightarrow x \leq 1 - \tau$ or $\tau \geq 1/2$ and $y \leq 1 - \tau$. Then,

$$|T| = z^{-1} \int_\mathbb{E} \int_\tau [T(P_\mathbb{E}, \theta_1, \tau) + T(P_\mathbb{E}, \theta_2, \tau) + T(P_\mathbb{E}, \theta_3, \tau) + T(P_\mathbb{E}, \theta_4, \tau)] d\mathbb{E} \, d\tau$$

$$= 2^{-1} [0 + 1/4 + 1/4 + 1/2] = 1/2.$$

Next, consider the entropy minimization policy. In this case $T(P_\mathbb{E}, A, \tau) = 1 \Leftrightarrow y \geq \tau$, while $T(P_\mathbb{E}, B, \tau) = 1 \Leftrightarrow y \leq 1 - \tau$. Now,

$$|T| = 2^{-1} [0 + 1/3 + 1/6 + 1/2] = 1/2.$$

Thus, we see that what entropy minimization loses in caution by committing to A, it offsets by a lack of commitment to B. On average, then, entropy minimization is just as cautious as entropy maximization.

While it might appear that caution cannot after all be used as an argument in favour of the Maximum Entropy Principle, such a conclusion would be too hasty. In fact, the Maximum Entropy Principle *is* most cautious where it matters, namely in the face of *risky* decisions, as we now see.

It is important to be cautious when a course of action should not be taken lightly, and as we have seen, the risk attached to a decision tends to be reflected in its trigger level. Thus, when deciding between meningococcal meningitis and 'flu, a high trigger level for 'flu indicates that the ensuing course of action (treatment of 'flu rather than meningitis) is risky. Similarly, when deciding between innocence and guilt in a criminal case, there is a high trigger level for guilt because the consequences of a mistaken judgement of guilt are considered dire. So when we consider the average caution of a policy for setting degrees of belief, it makes sense to focus on the decisions where caution is important, namely those sentences θ with a *high* trigger level τ_θ.

So let us assume in our two-dimensional example that $\tau \geq 1 - \epsilon$, where $0 \leq \epsilon \leq 1/2$ is small. This extra constraint means that now $z = 4 \times \epsilon/2 = 2\epsilon$. As before, $T(P_\mathbb{E}, \theta_1, \tau) = 1$ has measure 0, but now $T(P_\mathbb{E}, \theta_4, \tau) = 1$ has measure $\epsilon/2$ for both policies. In the entropy maximization case both $T(P_\mathbb{E}, A, \tau) = 1$ and $T(P_\mathbb{E}, B, \tau) = 1$ have measure $\epsilon^3/6$, and

$$|T| = (2\epsilon)^{-1}\left[0 + \epsilon^3/6 + \epsilon^3/6 + \epsilon/2\right] = (1 + 2\epsilon^2/3)/4.$$

On the other hand, in the entropy minimization case $T(P_\mathbb{E}, A, \tau) = 1$ has measure $\epsilon/2 - 1/6[1 - (1 - \epsilon)^3]$, while $T(P_\mathbb{E}, B, \tau) = 1$ has measure $\epsilon^3/6$, and

$$|T| = (2\epsilon)^{-1}\left[0 + \epsilon/2 - 1/6[1 - (1 - \epsilon)^3] + \epsilon^3/6 + \epsilon/2\right] = (1 + \epsilon)/4.$$

Thus, entropy maximization offers the smaller average proportion of triggers: if $\epsilon = 1/2$, then entropy minimization triggers in about 30% more cases, while if $\epsilon = 1/10$, it triggers in about 10% more cases.

In sum, it appears that one might, after all, appeal to caution to make a case for objective Bayesianism. Being close to the equivocator is *on average the more cautious policy when it comes to risky decisions*. This caution is explained by the fact that the more middling one's degrees of belief, the smaller the proportion of triggered decisions on average, when trigger levels are high.

Note that this line of justification appeals to the idea of trigger levels but is independent of the particular decision theory or method used to derive those trigger levels. It is merely assumed that, other things being equal, higher trigger levels correspond to higher risks attached to erroneous triggering: holding fixed the evidence, the range of options, and so on, if $\tau_\theta > \tau_\varphi$, then the consequences of taking the action triggered by $P(\theta) \geq \tau_\theta$, when θ turns out false, are more dire than the consequences of taking the action triggered by $P(\varphi) \geq \tau_\varphi$, when φ turns out false.

It remains to say something about why the measure of distance from the equivocator $P_=$ is taken to be divergence rather than, say, Euclidean distance.

If, in calculating the average proportion of triggers $|T|$, one takes averages using simple volumes (Lebesgue measure) then minimization of Euclidean distance from the equivocator appears to be optimal; but of course one might average in different ways to justify different measures of distance. Consequently, the appeal to minimization of the average proportion of triggers is best thought of as a suggestive rather than rigorous justification of the Equivocation norm, merely a sketch of how the idea of caution might be extended to dimensions higher than two. The appeal to divergence needs more motivation.

There are two main ways in which one can justify the appeal to divergence. The first consideration is that divergence is the concept of distance that has become entrenched in information theory and information geometry, two fields that need to consider distance between probability functions. In information theory, it is natural to interpret the distance of Q from P as the amount of information contained in Q that is not in P. Hobson (1971, pp. 36–9) argued that, under certain assumptions, the latter quantity should be measured by the divergence of Q from P; see also Paris (1994, pp. 121–2). This concept of distance has stuck, despite it being less attractive than Euclidean distance from a mathematical point of view (recall that divergence is not a distance function in the strict mathematical sense). Moreover, in information geometry cross-entropy divergence has been shown to be a very natural notion of distance for perhaps the most basic family of probability distributions, the exponential family (Amari and Nagaoka 1993, §3.5). From a geometric perspective, objective Bayesianism recommends adopting the projection of $P_=$ onto $\mathbb{E}$ as a belief function, with divergence minimization acting as the projection operation. So from the point of view both of information theory and of information geometry, this divergence function is well-established as a measure of distance between probability functions.

The second way in which one might try to justify the appeal to divergence is through betting behaviour. Suppose the agent bets on the atomic states $\omega \in \Omega$ with betting quotients $P(\omega)$ satisfying the Probability norm. Suppose too that the agent will lose $-\log P(\omega^*)$, where ω^* is the atomic state that turns out to obtain. Given the Calibration norm, it might seem prudent to choose $P \in \mathbb{E}$ that minimizes the agent's worst-case expected loss—this is called the *robust Bayes* choice of P. Since $\mathbb{E}$ is convex, it turns out that, under some rather general conditions, the robust Bayes choice will be the probability function that has minimum divergence from the equivocator, if there is such a function at all (Topsøe 1979; Grünwald and Dawid 2004). Hence, minimizing divergence from the equivocator is provably the most cautious policy in this precise sense.

Note, though, that this justification assumes a logarithmic loss function. Now, in any particular decision problem a different loss function may in fact be in operation. This might suggest that the justification has rather limited scope. However, two considerations should be borne in mind. First, degrees of belief are typically ascertained in the absence of any knowledge of the loss function in operation. Second, the same degrees of belief may be applied to several different decision problems: the agent's current belief function must last her until her body of evidence changes and might in the meantime be put to use in a range of decisions. Hence, there is scope for justifying the appeal to divergence if logarithmic loss can be shown to be an appropriate *default* loss function—one that should apply in the absence of evidence about the true loss function, or in the presence only of limited evidence of the loss function (e.g. in the presence of evidence of the loss functions of some but not all of the decisions that might be taken with the current belief function). Now, logarithmic loss is in fact often taken to be a default loss function (see, e.g. Good 1952, §8; Rissanen 1987). There are several rather inconclusive reasons for this: for example, this loss function corresponds very nicely to certain coding games, and it generates a strictly proper scoring rule, which is desirable in itself. But there are also two more decisive arguments, which we briefly consider.

One argument for logarithmic loss as a default appeals to betting again: the *Kelly gambling* or *proportional gambling* set-up (Kelly 1956; Cover and Thomas 1991, chapter 6). In this betting set-up, odds are announced in advance—a return of o_i pounds for each pound bet on ω_i if ω_i turns out true—and the agent bets $q_i W$ on each ω_i, where q_i is her betting quotient for ω_i and W is her total wealth. This betting set-up is rather natural: horse racing works like this (though one may be advised not to bet one's total wealth on a horse race!). It turns out that if bets on the ω_i are made repeatedly and repeated outcomes of the ω_i are iid with respect to chance P^*, then the agent should set her betting quotients by the function in $\mathbb{E}$ that has minimum divergence from the equivocator in order to maximize the worst-case expected growth rate of her wealth (Grünwald 2000, §5.1). Hence, divergence turns out to be just the right measure of closeness to minimize worst-case expected loss rate in this rather plausible betting set-up. To the extent that the long-term application of degrees of belief can by default be modelled by this set-up, logarithmic loss is a good default loss function and divergence is a good default measure of distance. (Interestingly, this Kelly gambling set-up can also be used to motivate the Calibration norm: the expected growth rate of the agent's wealth is maximal just when the agent's belief function matches the chance function (Cover and Thomas 1991, Theorem 6.1.2).)

Yet one might not be entirely satisfied by the Kelly gambling argument. One reason is that its conclusions depend on betting one's entire wealth on each race, which seems rash and unrealistic. Another is that the argument is restricted to repeated trials (e.g. a repeated horse race) and one might seek something more general. In fact, something more general is possible with a second kind of argument—an argument by derivation whose assumptions are so slender that it is hard to heap aspersions on them. Denote the loss one incurs by adopting belief function P when state ω is the case by $L(\omega, P)$. (Here the decision problem involves deciding on a probability function P; in such cases a loss function L is often called a *scoring rule*.) Assume the following of the default loss function L:

L1: By default, $L(\omega, P) = 0$ if $P(\omega) = 1$.
L2: By default, loss strictly increases as $P(\omega)$ decreases from 1 towards 0.
L3: By default, loss $L(\omega, P)$ depends only on $P(\omega)$, not on $P(\omega')$, for $\omega' \neq \omega$.

L4: By default, losses are presumed additive when the language is composed of independent sublanguages: if $\mathcal{L} = \mathcal{L}_1 \cup \mathcal{L}_2$ and $\mathcal{L}_1 \perp\!\!\!\perp_P \mathcal{L}_2$, then $L_{\mathcal{L}}(\omega_1 \wedge \omega_2, P) = L_{\mathcal{L}_1}(\omega_1, P|_{\mathcal{L}_1}) + L_{\mathcal{L}_2}(\omega_2, P|_{\mathcal{L}_2})$. (Here ω_1 and ω_2 are atomic states of $\mathcal{L}_1$ and $\mathcal{L}_2$, respectively.)

If these conditions are to hold for any finite propositional language $\mathcal{L}$ and any probability function P defined on this language, then the default loss function L must be logarithmic loss, $L(\omega, P) = -\log_b P(\omega)$ to some base b. One can see this as follows. L3 implies that $L(\omega, P) = f(P(\omega))$, some function of $P(\omega)$. According to L4, $f(P(\omega_1 \wedge \omega_2)) = f(P(\omega_1)P(\omega_2)) = f(P(\omega_1)) + f(P(\omega_2))$. But the negative logarithm on $(0, 1]$ is characterizable in terms of this additivity, together with the condition that $L(\omega, P) \geq 0$ which is implied by L1–2 (see, e.g. Aczél and Daróczy 1975, Theorem 0.2.5). (Accordingly, L1–2 can be weakened to the condition that $L(\omega, P) \geq 0$.)

So one can justify the Equivocation norm on the grounds that (i) minimising cross-entropy divergence from the equivocator minimises worst-case expected loss when the loss function is logarithmic and (ii) it is natural to take loss as being logarithmic in the absence of knowledge of the true loss function. As with the other norms, this motivation of Equivocation appeals to betting considerations and the avoidance of loss.

In sum, the Equivocation norm can indeed be motivated by considerations to do with caution. Intuitively, if evidence only constrains the strength of the consultant's belief that her patient's cancer will recur to lie in the interval $[0.1, 0.4]$, then the consultant would be irrational to adopt a very low degree of belief because that could lead her to take unwarranted risky actions. Further consideration backs this intuition up: adopting extreme degrees of belief appears to be on average a less cautious policy with respect to risky decisions. So there is clearly something right about ruling out sufficiently extreme degrees of belief as irrational. But that means ruling in as rational those degrees of belief that are sufficiently equivocal. Our choice of equivocator and distance measure are very natural, given the experience of information theory and information geometry, and given the desirability of minimizing worst-case expected loss. But the objective Bayesian should not be wedded to cross-entropy divergence and maximizing entropy. In particular, knowing that one's degrees of belief will only be applied within the context of a particular non-logarithmic loss function would motivate a different distance measure. While in this book we focus for concreteness on divergence, other measures of distance remain an interesting topic for further research.

We see that just as successful calibration is something of an art, so too is successful equivocation. One can calibrate to the extent that one can identify what one grants and identify the constraints imposed by this evidence. Similarly, one can equivocate to the extent that one can identify the equivocator and identify those functions sufficiently close to the equivocator. Now for a finite propositional language there is no problem identifying the equivocator, though we see in §9.1 that this is not the case for certain uncountable domains. But it can be harder to identify those functions sufficiently close to the equivocator. It can be hard to identify the pragmatic constraints to say how close counts as 'sufficiently' close. It may also be hard to identify the right notion of distance from the equivocator. If a range of distance functions are plausible—induced, perhaps, by a range of plausible loss functions—then it would be natural to include any probability function deemed sufficiently equivocal with respect to *some* plausible distance function within one's set $\Downarrow\mathbb{E}$ of sufficiently equivocal functions. That calibration is an art does not mean that the Calibration norm should be abandoned; similarly, that equivocation is an art does not mean that the Equivocation norm should be abandoned.

Although the Equivocation norm encapsulates a need to be cautious, it should not be confused with the *Precautionary Principle* (see, e.g. Harremoës *et al.* 2002). This latter principle is construed in various ways: one formulation has it that one should not act if there is insufficient evidence that the act in question is harmless, another maintains that one should act to mitigate harm even in cases where there is inconclusive evidence that the act in question will really mitigate harm. Whatever the accepted formulation, however, the Precautionary Principle is at root a principle concerning how one should set what we have called *trigger levels* for action, rather than how one should apportion one's beliefs: one should have a high trigger level for an action that might increase harm, or, one should have a low trigger level for an action that might decrease harm. The Equivocation norm says that one should not strongly believe a proposition unless forced to by evidence, on the grounds that one would otherwise behave rashly. On the other hand, the Precautionary Principle says just that one should not behave rashly, in the sense that trigger levels for an action should be guided by what evidence there is concerning the consequences of the action for harm.

3.4.5 *Sensitivity analysis*

We have seen that an appeal to caution can be used to motivate the move from empirically based subjective Bayesianism to objective Bayesianism: the latter is more cautious on average with respect to risky decisions. We now consider a possible response. Arguably, there is an extension of empirically based subjective Bayesianism that is more cautious still than objective Bayesianism.

Suppose evidence constrains a degree of belief to an interval. Suppose too that empirically based subjective Bayesianism is adopted, so that an agent may choose any degree of belief within this interval. One may want to be very cautious and avoid taking a course of action if the decision to do so depends on the degree of belief chosen. This leads to the following modification of the decision rule: instead of embarking on a course of action if and only if one's own degree of belief triggers the action, embark on it just when every possible agent in the same situation would too, that is, if and only if the whole interval of possible degrees of belief is greater than the trigger level. This decision rule might be called the *sensitivity analysis* rule—a decision is only taken if it is not sensitive to subjective choice of prior probability.

Under this view, degrees of belief are partly a matter of subjective choice, but, once trigger levels are chosen, it is an objective matter as to whether a decision will be triggered. In two dimensions, this decision procedure exhibits the same average caution as objective Bayesianism over risky decisions. However, in higher dimensions it will be more cautious in general. Recall the example of Fig. 3.3: here the Maximum Entropy Principle triggers both decisions, the Minimum Entropy Principle triggers one decision, while the sensitivity analysis decision procedure triggers neither.

Thus, it seems that by appealing to caution the objective Bayesian is shooting herself in the foot. While such an appeal favours objective Bayesianism over empirically based subjective Bayesianism as normally construed, it also favours the sensitivity analysis modification of empirically based subjective Bayesianism over objective Bayesianism.

However, the sensitivity analysis approach is conceptually rather unattractive. This is because it divorces the connection between belief and action: under the sensitivity analysis approach one's degrees of belief have no bearing on whether one decides to take a course of action. It matters not a fig the extent to which one believes a patient has meningitis, because the decision as to what treatment to give is based on the *range* of permissible beliefs one might adopt, not on one's own actual beliefs. Arguably, this is an unacceptable consequence of the sensitivity analysis approach which more than offsets its merit with respect to considerations of caution.

But perhaps there is some way of putting the sensitivity analysis approach on firmer footing. Perhaps, one can retain its caution while re-establishing the link between belief and action. We investigate two possible strategies for salvaging this approach.

3.4.6 *Higher-order beliefs*

The sensitivity analysis approach bases decisions on the range of beliefs that an agent might adopt, rather than on the agent's own beliefs. If one wants to retain this decision mechanism but also to insist that an agent's decisions be made on the basis of her own beliefs then one must reinterpret her beliefs as somehow encapsulating this whole range.

Suppose for example that empirical constraints force $P(\theta) \in [0.7, 0.8]$. Under the sensitivity analysis approach, the agent is free to choose whichever degree of belief she likes from this interval, but a decision on A will only be triggered if the whole interval triggers, that is if 0.7 is greater than the trigger level. Consider an alternative viewpoint—a Bayesian who is uncertain as to which degree of belief x to adopt from within this interval. In the face of this uncertainty, higher-order degrees of belief, such as $P(x \in [0.7, 0.75])$, become relevant. Indeed, the agent may form a prior belief distribution over x, and base her decision for θ on various characteristics of this distribution. One decision rule, for instance, involves triggering θ if $P(x \geq \tau_\theta) = 1$.

This alternative viewpoint yields a type of empirically based subjective Bayesianism: an agent's degrees of belief are constrained just by evidence. It is also compatible with cautious decision rules, such as that exemplified above. Moreover, by admitting higher-order degrees of belief it reinstates the link between belief and action: decisions are triggered by features of these higher-order beliefs. Thus, this approach appears to combine the best of all worlds—perhaps for that reason higher-order beliefs are popular in Bayesian statistics.

But all is not plain sailing for higher-order degrees of belief. A decision rule, such as that given above, is only cautious under some priors over x. If x is uniformly distributed over $[0.7, 0.8]$, then $P(x \geq \tau_\theta) = 1$ iff $0.7 \geq \tau_\theta$, the same cautious decision rule as the sensitivity analysis approach. On the other hand, if the probability of x is concentrated on the maximal point, $P(x = 0.8) = 1$, then the decision on θ triggers just when $0.8 \geq \tau_\theta$—in this case the decision rule is considerably less cautious, and in particular less cautious than the Maximum Entropy Principle. Now, the empirically based subjective Bayesian cannot advocate setting one prior rather than another, since there is no extra empirical evidence to constrain choice of prior. Indeed, the agent is free to choose any prior she wishes, and if she sets $P(x = 0.8) = 1$, she is far from cautious. So suggesting that the agent forms a prior over priors only defers the problem, leading to a vicious regress.

Thus, we see that while the higher-order belief approach is compatible with cautious decision rules, it is also compatible with rash decision rules. It certainly cannot be argued that this approach is any more cautious than the objective Bayesian method. Higher-order beliefs do not, after all, lead to the salvation of sensitivity analysis.

3.4.7 Interval-valued beliefs

There is another way one might try to salvage the cautiousness of sensitivity analysis. Again, this involves reinterpreting an agent's beliefs as encapsulating the whole range of empirically constrained values. But this time, rather than invoking uncertainty as to which degree of belief to adopt, one instead rejects the Bayesian idea that an agent's partial beliefs are numerical point-valued degrees of belief, that is, probabilities (Walley 1991, §1.1.5). Under this approach, introduced in §3.2, an agent's partial belief in θ is identified with the whole interval yielded by empirical constraints, $bel(\theta) = [0.7, 0.8]$ in our example.[9]

---

[9] Kyburg Jr (2003) provides a recent discussion of this strategy. See §7.2 and Kyburg Jr and Teng (2001) for the formal details of Kyburg's approach, which he called *evidential probability*. Kyburg Jr and Teng (1999) argue that the interval approach performs better than the subjective Bayesian approach in the short run in a betting set-up.

Note that the interval approach is sometimes motivated by the following kind of argument: objective Bayesians say that a degree of belief $\frac{1}{2}$ in θ is equivocal, but it is nothing of the sort since it commits to the probability of θ being $\frac{1}{2}$; only an interval-valued belief of $[0, 1]$ is truly equivocal (Walley 1991, §1.1.4.b; de Cooman and Miranda 2007). This argument rests on a simple misunderstanding. Degree of belief $\frac{1}{2}$ commits to the probability of θ being $\frac{1}{2}$ only in the sense of epistemic probability (rational degree of belief), not in the sense of physical

Motivation

Interval-valued beliefs offer a very appealing reconstruction of the sensitivity analysis approach. A natural decision rule proceeds thus: trigger a course of events on θ iff the agent's partial belief in θ is entirely above the trigger level for θ, in our example, iff $[0.7, 0.8] \geq \tau_\theta$, that is, iff $0.7 \geq \tau_\theta$. Not only does the resulting framework capture the cautiousness of the sensitivity analysis approach, but also ties the triggering of a decision to the agent's own partial belief, rather than the beliefs of other possible agents.

The crunch point is this. The partial belief approach appears to be superior to objective Bayesianism with respect to caution; does any such gain outweigh any difficulties that accompany the rejection of point-valued degrees of belief in favour of interval-valued beliefs?

Perhaps not. First of all, there are qualifications to be made about the cautiousness of interval-valued beliefs that diminish their supposed superiority. Second, the problems that accompany interval-valued beliefs arguably outweigh any remaining benefit in terms of caution.

First to the qualifications. One way of generating interval-valued beliefs runs as follows (Kyburg Jr and Teng 2001, §11.3). Sample an attribute A from a population; say the sample frequency is 0.75; under certain assumptions about the sampling mechanism and the population, one might form a confidence interval, say $[0.7, 0.8]$ for the population frequency; set one's partial belief in an individual having attribute A to this confidence interval. The key problem with this approach is that the confidence interval will depend upon the chosen confidence level as well as the sampling assumptions—thus, the end points of the interval are somewhat arbitrary. But the decision procedure depends crucially on the end points: a 95% confidence interval $[0.7, 0.8]$ may trigger a course of action for A while a 99.9% confidence interval $[0.5, 1]$ may fail to trigger the action.

But there may be no non-ad hoc way of determining a suitable confidence interval and so this approach must be abandoned if one wants an objective decision procedure. Perhaps the best alternative strategy is just to set one's partial belief in A to the sample frequency 0.75—this is, after all, the most probable candidate for the population frequency. But then the belief is not interval-valued after all, it is point-valued. Thus, the interval-valued approach loses its edge. (Arguably, this is a qualification to the interval-valued belief approach rather than a reason to dismiss it altogether: one can still adopt an interval-valued belief in certain circumstances, for example if there are two samples with sample frequencies 0.75 and 0.77, respectively then, as argued in §3.3, it is natural to view the whole interval $[0.75, 0.77]$ as a candidate for partial belief. Moreover, the confidence interval idea may be salvageable even where there is no most appropriate confidence interval, by appealing to higher-order probabilities: one may believe to degree 0.95 that one's partial belief should be in $[0.7, 0.8]$ and believe to degree 0.999 that it should be in $[0.5, 1]$ (§10.2).)

probability—it implies nothing about the chance of θ. Degree of belief $\frac{1}{2}$ *does* equivocate between θ and $\neg\theta$ in the sense that neither alternative is believed more than the other. It also equivocates between chances in the trivial sense that it says nothing about chance.

We have seen then that the advantages of interval-valued beliefs with respect to caution are somewhat diminished. Next, we turn to the problems that accompany interval-valued beliefs. As noted above, it is not empirical evidence that adjudicates between the two approaches; the problems with interval-valued beliefs are pragmatic and conceptual.

From the pragmatic point of view, it is harder to obtain and work with interval-valued beliefs than point-valued beliefs. Roughly speaking, it is twice as hard, since there are twice as many numbers to have to deal with: to each point-valued degree of belief there are the two end points of the corresponding interval-valued belief. Accordingly one might reason something like this: the formalism of point-valued degrees of belief offers a first approximation to how one should reason; the formalism of interval-valued beliefs offers a second approximation: the first approximation tends to be easier to use in practice and there is little to be gained (in terms of caution) by using the second approximation; thus, one should use the first approximation by default. But such a view assumes that interval-valued beliefs are conceptually on a par with point-valued degrees of belief. There are reasons for doubting this perspective, as we now see.

From the conceptual point of view, the interval approach—like the sensitivity analysis approach—weakens the link between belief and decision. The decisions we take depend on the strengths of our beliefs, in the sense that believing a proposition sufficiently strongly will lead to one action, otherwise another will be taken. More precisely, it is plausible that if a decision depends on the strength of one's belief in θ, then there is a threshold τ such that if the strength of belief in θ passes this threshold, one will perform action a; otherwise, if it fails to meet this threshold but passes a lower threshold τ', action a' will be taken; and so on. The thresholds $\tau, \tau', \ldots$ partition the unit interval into subintervals, each of which will trigger a different action, ensuring that some course of action (or inaction) $a, a', \ldots$ will be taken. This is compatible with point-valued beliefs, where strengths of belief are linearly ordered. But it is not compatible with interval-valued beliefs. Suppose, for example, that a doctor has to decide how to treat a patient whose symptoms fail to distinguish between a dangerous illness θ that would require urgent treatment and a more benign alternative $\neg \theta$ that needs little or no treatment at at all (meningococcal meningitis versus 'flu, for example). If the doctor has interval-valued belief $[0,1]$ in θ and $[0,1]$ in $\neg\theta$, representing complete ignorance concerning θ, then there is no threshold τ that will differentiate as to which action should be taken. On the other hand, if the doctor has point-valued belief $\frac{1}{2}$ in θ and $\frac{1}{2}$ in $\neg\theta$, representing total equivocation between θ and $\neg\theta$, then the value $\frac{1}{2}$ can be compared with a threshold τ (which is likely be lower than $1/2$ in this example, given the risks attached to the dangerous illness) allowing a decision as to which action should be taken. In short, if action is to depend on strength of belief and suitable thresholds, then strengths of belief need to be linearly ordered, as they are with the betting interpretation of §3.1.

There is a second conceptual worry, alluded to in §3.2. One of the key points in favour of the Bayesian approach is that an agent's partial belief in E is interpretable as a betting quotient, 'the rate p at which he would be ready to exchange the possession of an arbitrary sum S (positive or negative) dependent on the occurrence of a given event E, for the possession of the sum pS' (de Finetti 1937, p. 62). One cannot simply identify an interval-valued partial belief with a betting quotient—a betting quotient is a single number but an interval is a set of numbers. One might try, as Borel (1943, §3.8) did, to interpret the interval as a *set* of acceptable betting quotients; see also Smith (1961) and Walley (1991, §1.6.3). To do so, one must adopt a different betting set-up to that of the Bayesian, one without the requirement that the agent buys and sells bets at the same rate, that is, one in which S must be positive rather than either positive or negative. (Otherwise, if the agent has more than one betting quotient in the same outcome, then she can simply be Dutch-booked—a set of bets can be found that forces her to lose money whatever happens.) This different betting set-up does seem compelling when we consider real bets: after all we do not, in fact, make bets by agreeing to buy and sell at the same rate. However, this move is less than compelling as an interpretation of rational strength of belief. That I am only prepared to bet on θ at odds that I regard as extremely favourable says nothing about how strongly I believe θ, let alone how strongly I ought to believe θ. But if it can be shown that, were I to offer betting quotient q both for buying and selling bets on θ, q should satisfy some constraint χ, then plausibly the strength of my belief in θ should also satisfy that constraint. Now this kind of argument appeals to the standard betting set-up of §3.1, not the modification required by the interval approach. So while positive stakes may be better for modelling real bets, buying and selling at the same rate is better for explicating the concept of rational degree of belief.

The proponent of intervals may respond here that the Bayesian link between partial beliefs and betting quotients is less attractive than one might think. One objection to the Bayesian betting set-up is that human agents cannot always evaluate their betting quotients in terms of unique real numbers—human beliefs are simply not so precise (see, e.g. Kyburg Jr 1968). Another is that it is rather impractical to elicit degrees of belief using a series of bets, since this is a time-consuming operation, and in any case people are often either reluctant to bet at all or happy to lose money whatever happens (§3.1). These points are well made, but by the by in our context because they only trouble the subjective and empirically based versions of Bayesianism, not objective Bayesianism. Under objective Bayesianism, agents do not need to search their souls for real numbers to attach to beliefs, nor is the betting set-up required to measure those degrees of belief. Degrees of belief are determined by the three norms and they are measured by techniques for determining equivocal degrees of belief such as the Maximum Entropy Principle; human agents and artificial agents alike can use computing power to work out the extent they ought to believe various outcomes. Thus, for objective Bayesianism the betting interpretation is only important for the *meaning* it gives to degrees of belief. The fact that the betting set-up is simplistic, or an idealization, or impractical, is neither here nor there—the objective Bayesian does not go on to invoke the betting set-up, as Ramsey did, as an elicitation or measurement tool.

One might argue that interval-valued beliefs are superior to point-valued beliefs because they can represent the amount of evidence that grounds a belief. Under the objective Bayesian account, a degree of belief 0.5 that it will snow in Canterbury today may be based on total ignorance, or it may be based on good evidence, for example, the knowledge that the frequency of it snowing in Canterbury on a day like today is $\frac{1}{2}$. In contrast, the interval-valued approach distinguishes between these two cases: while in the latter case the belief might have value 0.5, in the former case the belief would have value $[0,1]$. In a sense, then, interval-valued beliefs tell us about the evidence on which they are based; this is supposed to be an advantage of the interval-valued approach (see, e.g. Walley 1991, §1.1.4.a; de Cooman and Miranda 2007, §1; Sturgeon 2008, §7).

But this points to a third conceptual problem with the interval-valued approach, not an advantage. The problem is this. The question 'how much evidence does the agent have that pertains to it snowing in Canterbury today?' is a question about evidence, not belief. Consequently, it should not be the model of belief that answers this question; it should be the evidential component of the agent's epistemic state. But on the interval-valued approach, it is the belief itself that is used to answer the question (and typically there is no separate model of the agent's evidence). Thus on this approach, the belief model conflates concerns to do with evidence and belief. On the other hand, the objective Bayesian approach maintains a sharp distinction between evidence and belief: an agent has evidence which is then used to constrain her choice of degrees of belief; the former component contains information about what the agent grants, while the latter component contains information about the strengths of the agent's beliefs; neither need be used to answer questions about the other. The interval-valued approach, then, muddles issues to do with evidence and belief, while the objective Bayesian approach keeps them apart. But the goal of both approaches is to model a rational agent's epistemic state, and this requires a sharp distinction between evidence and belief. It is a fact of the matter that some of our full beliefs are of a higher grade than others, not up for negotiation in the current operating context. I have a full belief that a point chosen at random in a ball will not be its centre; I also have a full belief that I am alive; the former belief may be quite negotiable but the latter not—it is the distinction between believing and granting that accounts for this difference. Hence, the objective Bayesian approach, by maintaining a proper distinction between evidence and belief, offers a better model of an agent's epistemic state.

To conclude, any gains that interval-valued beliefs offer in terms of caution are arguably offset by the conceptual as well as pragmatic advantages of point-valued beliefs.

3.5 Radical subjectivism

One kind of subjectivist objects to objective Bayesianism on the grounds that the Equivocation norm goes well beyond the usual tenets of subjectivism; we have seen though that this norm does admit some motivation. But perhaps as awkward for the objective Bayesian is the subjectivist who *does not* object to objective Bayesian assignments of degrees of belief. Such a subjectivist may say something like this: the objective Bayesian satisfies all the requirements of subjective Bayesianism—give or take a few discrepancies when updating (Chapter 4)—so there is nothing to object to; if her degrees of belief are always maximally equivocal, that is fine since subjectivism is a broad church.

The question then arises as to the status of the norms of objective Bayesianism—are they a voluntary code of conduct, or is an agent somehow compelled to abide by them?

This question points to a deeper worry that affects objective and subjective Bayesianism alike. What compulsion is there to abide by *any* norm of rational belief? The complaint goes: you may say that my degrees of belief should be such and such, but I do not consider myself to be subject to your rules. This leads to what might be called *radical subjectivism* or *radical voluntarism*, the view that rules of rational belief ascription express individual inclinations or value judgements and cannot be imposed on others.[10] Under this position, objective and subjective Bayesianism are just two stances one might take as to how to set one's degrees of belief; neither can be considered right or wrong. A radical subjectivist might agree that if you value avoiding sure loss, then you should abide by the Probability norm, for otherwise you are prone to a Dutch book. But, the radical subjectivist might say, I do not take avoiding sure loss to be paramount even in the synchronic case, let alone in the diachronic case. Similarly, the radical subjectivist may disavow minimizing expected or long-run loss (Calibration) and may not mind exposing himself to unnecessary risks (Equivocation).

One might be inclined to bite the bullet here and accept radical subjectivism, safe in the knowledge that others are in the same camp as oneself. Thus, one might express the norms of rational belief in terms of means–ends rationality: *if* your ends are the minimization of loss (sure loss, long-run loss, worst-case expected loss), then you had better be a good objective Bayesian. If not, not. But as a matter of fact most of us have these values or ends, so we should abide by these norms. Then, objective Bayesianism speaks to those who take the required stance; others may be perfectly rational to flout its rules.

But one can do better than this. When we talk of degrees of belief, we talk of something practical: the strengths of our beliefs guide our behaviour. We use our degrees of belief to make decisions and to draw various inferences (e.g. to make predictions) and these are practical uses. Now, the ends or values connected with objective Bayesianism come part and parcel with these uses. When we make decisions we want them to be good decisions—we do not want to take unnecessary

---

[10] Note that van Fraassen's voluntarism is less radical inasmuch as it holds that while rational norms incorporate value judgements, they are open to some sort of debate. See Fraassen (1984, 1989, pp. 174–5, 2002). For a discussion of radical subjectivism see *Episteme: A Journal of Social Epistemology* 4(1), 2007, a special issue on epistemic relativism.

risks, for instance. When we make predictions we want them to be correct, not disjoint from reality. So merely by talking of degrees of belief, we buy into these values. The ends are in the meaning of belief, they are not optional.

The radical subjectivist might take issue with this by saying: when I say 'degrees of belief' I do not buy into these values; I do not care whether or not I make good decisions on the basis of my 'degrees of belief'; I talk about them in isolation from their consequences. Fine, but then the radical subjectivist is talking about something else—what we might call 'degrees of idle-belief', not the degrees of belief that do unavoidably ground his actions. The Bayesian may then respond by admitting that her norms do not govern the radical subjectivist's degrees of idle-belief, but insisting that they do apply to his degrees of belief.

In that sense, then, the norms of objective Bayesianism are compulsory. You can use the terminology 'degrees of belief' to refer to anything you wish, but what I call your 'degrees of belief' had better abide by Probability, Calibration, and Equivocation.

Consider an analogy: constructing a *boat* rather than constructing *degrees of belief*. Boats are used to travel along the surface of water, and floating is clearly part and parcel of the proper functioning of a boat. A boat is still a boat if it has been holed and sinks, so floating is not an essential property of a boat, but if you take it upon yourself to build a boat, you should make sure it floats on water. This *Floating* norm is thus imposed on the boat-building process rather than the boat itself. The radical boat-builder might not buy into this norm, saying: I am not interested in travelling on water when I build my boat. But then the radical boat-builder is not building a boat in the normal sense of the word. Rather, he is building something that might be called an 'art boat' (if it is being built for its aesthetic properties) or an 'idle boat' (if it is to have no uses at all). Determining degrees of belief is like building a boat, inasmuch as once one has decided on the thing one is constructing, one is bound to construct it in such a way that it has its proper function, otherwise one fails in one's task.

In sum, one can after all defend norms of rational belief—such as those that characterize objective Bayesianism—from the clutches of the radical subjectivist.

4

Updating

The goal of this chapter is to respond to a criticism of objective Bayesianism that concerns the updating of degrees of belief. Objective Bayesianism has often been dismissed on account of differences between its form of updating and Bayesian conditionalization (§4.1)—such differences are typically taken to count against objective Bayesianism. But we see in §4.2 that such differences in fact count against Bayesian conditionalization and that conditionalization can only safely be used where its results agree with those of objective Bayesian updating. Section 4.3 rehearses several well-known criticisms of conditionalization. We then go on to dismiss two justifications of conditionalization. Section 4.4 considers the dynamic Dutch book argument for conditionalization and offers a reductio against dynamic Dutch book arguments in general. Section 4.5 examines the argument for conditionalization that appeals to stability of conditional probabilities, and shows that this argument fails to offer a justification of conditionalization where it disagrees with objective Bayesian updating. Overall, conditionalization compares poorly with objective Bayesian updating (§4.6).

As before, we consider a finite propositional language $\mathcal{L}$. For simplicity of exposition we also focus on the case in which evidence $\mathcal{E}$ imposes a consistent set χ of affine constraints.[1] This ensures that $\mathbb{P}_\chi \stackrel{df}{=} \{P \in \mathbb{P}_\mathcal{L} : P \text{ satisfies } \chi\}$ is non-empty, closed, and convex; hence, $\mathbb{E} = \mathbb{P}_\chi$ and $\downarrow\mathbb{E}$ is a singleton. For simplicity, in this chapter we also identify 'sufficiently close' and 'closest', $\Downarrow\mathbb{E} = \downarrow\mathbb{E}$. As usual we take distance between probability functions to be explicated by divergence. Under these restrictions the objective Bayesian recommendation is to set one's belief function $P_\mathcal{E}$ to be the function in $\mathbb{P}_\chi$ that has maximum entropy.

4.1 Objective and subjective Bayesian updating

Under objective Bayesianism, updating in the light of new evidence is rather straightforward. If an agent initially takes for granted propositions in $\mathcal{E}$, but this then changes to $\mathcal{E}'$, then her degrees of belief should change correspondingly from $P_\mathcal{E}$ to $P_{\mathcal{E}'}$. On a finite propositional language, when granting $\mathcal{E}$ the agent should select the $P_\mathcal{E} \in \mathbb{E}$ that maximizes entropy, and then when granting $\mathcal{E}'$ she should select the $P_{\mathcal{E}'} \in \mathbb{E}'$ that maximizes entropy. Degrees of belief track

---

[1] An *affine* constraint is satisfied by every probability function $P_\lambda = (1-\lambda)P_0 + \lambda P_1$ that lies on a line through any two probability functions P_0, P_1 that satisfy the constraint. On a finite space, affine constraints take the form of sets of expectation constraints, $\sum_\omega P(\omega)f(\omega) = c$, for real-valued function f and real number c. Maximum entropy methods were originally applied solely to affine constraints for computational reasons, although this restriction may be relaxed.

the agent's evidence; since the agent's evidence $\mathcal{E}$ strongly constrains (under the assumptions of this chapter, fully constrains) her degrees of belief, changes to this evidence also strongly constrain changes to her degrees of belief.

As mentioned in §2.2, subjective Bayesians advocate the Probability norm, and often advocate the Calibration norm, but do not advocate Equivocation. Probability and Calibration on their own only weakly constrain degrees of belief: granting $\mathcal{E}$, choose some $P \in \mathbb{E}$. Subject to just these constraints, one's degrees of belief are permitted to vary wildly under small (or no) changes to one's evidence $\mathcal{E}$. Accordingly, subjective Bayesians impose a further principle in order to more strongly constrain updated degrees of belief. This is the principle of Bayesian Conditionalization: if on granting $\mathcal{E}$ your degrees of belief are represented by P, and if your evidence changes to $\mathcal{E}' = \mathcal{E} \cup \{\varepsilon\}$, where ε is a sentence of your language such that $P(\varepsilon) > 0$, then your new degrees of belief should be represented by $P' = P(\cdot|\varepsilon)$, that is, by your old degrees of belief conditional on the new evidence. This principle does the job very well—provided the conditions on ε are satisfied, P' is uniquely determined.[2]

Unsurprisingly perhaps, these two different approaches to updating can lead to different recommendations as to which degrees of belief to adopt in the light of new evidence. A standard example proceeds as follows. Suppose there are two agents, Olive (who abides by objective Bayesian norms) and Sylvester (who follows the subjectivist path). They both know $\mathcal{E}$, namely that an experiment is being performed which has three possible outcomes. We represent this space of outcomes by $\Omega = \{\omega_1, \omega_2, \omega_3\}$.[3] By Equivocation, Olive sets $P^O_\mathcal{E}(\omega_1) = P^O_\mathcal{E}(\omega_2) = P^O_\mathcal{E}(\omega_3) = 1/3$. Sylvester can choose any belief distribution he likes, but we suppose that his beliefs match Olive's, $P^S = P^O_\mathcal{E}$. Now, both agents are presented with new information ε, which says that $1 \times P^*(\omega_1) + 2 \times P^*(\omega_2) + 3 \times P^*(\omega_3) = x$, that is, with respect to physical probability P^* the expectation E of the outcome takes some particular value x in the interval $[1, 3]$. In this case, $\mathbb{P}_{\chi'} = \{P : 1 \times P(\omega_1) + 2 \times P(\omega_2) + 3 \times P(\omega_3) = x\}$ is the set of probability distributions

---

[2]Conditionalization is called a *conservative* updating rule since it tries to conserve prior degrees of belief. In contrast, objective Bayesian updating is *foundational* in the sense that objective Bayesian degrees of belief track their foundations, namely the agent's evidence. Historically, the split between objective and subjective Bayesian updating has not always been clear cut—some objective Bayesians have advocated conditionalization (Jaynes 2003) (as have proponents of logical probability such as Carnap), while some subjectivists have only endorsed conditionalization in a limited way (Howson 1997). The points made in the following sections remain: objective Bayesians *need not* advocate conditionalization, and indeed *should not* where conditionalization disagrees with the objective Bayesian approach presented above.

[3]Here, let A_1, A_2, A_3 be the elementary sentences of the agent's language with A_i expressing the proposition that outcome i occurs. Since these sentences are mutually exclusive and exhaustive, only three of the eight atomic states of the form $\pm A_1 \wedge \pm A_2 \wedge \pm A_3$ are genuine possibilities. For simplicity, we restrict attention to these three possibilities, taking ω_1 to be $A_1 \wedge \neg A_2 \wedge \neg A_3$, ω_2 to be $\neg A_1 \wedge A_2 \wedge \neg A_3$, and ω_3 to be $\neg A_1 \wedge \neg A_2 \wedge A_3$, and ignoring the other five states. Since ω_i and A_i are equivalent, they can be used interchangeably. In general, if $\mathcal{L} = \{A_1, \ldots, A_n\}$, where the A_i are mutually exclusive and exhaustive we sometimes simply take $\Omega = \{A_1, \ldots, A_n\}$ and take it as understood that each A_i can either be viewed as an elementary sentence or an atomic state.

that yield expected outcome $E = x$; this is non-empty, closed, and convex, so plausibly $\mathbb{E}' = \mathbb{P}_{\chi'}$. Olive then chooses the distribution in $\mathbb{E}'$ that has maximum entropy as her new belief function $P_{\mathcal{E}'}^O$. Now Sylvester cannot update at all in this situation, because ε is not in the outcome space. But we can ask what would happen were the outcome space enlarged to incorporate ε, and were Sylvester then to update his degrees of belief by conditionalizing on ε. It turns out that if Sylvester's updated degrees of belief are to match Olive's, he would have to have given prior probability 1 to the outcome $E = 2$ (Shimony 1973). This seems ridiculous, since it amounts to prior certainty that the initial probability will not be revised. Of course, Sylvester is a subjectivist and is rationally permitted to believe $E = 2$ to any degree he wishes, and if he were less than certain that $E = 2$, then his degrees of belief would differ from Olive's after updating.

The Judy Benjamin problem can be interpreted as providing another example of a disagreement between the objective and subjective Bayesian versions of updating (van Fraassen 1981, 1987, 1989; van Fraassen et al. 1986). Private Judy Benjamin is dropped into a swampy area which is divided into four quadrants each occupied by a different company: Red HQ (R_1), Red 2nd (R_2), Blue HQ (B_1), Blue 2nd (B_2). Benjamin does not know which zone she is in, and initially equivocates, setting $P_{\mathcal{E}}(R_1) = P_{\mathcal{E}}(R_2) = P_{\mathcal{E}}(B_1) = P_{\mathcal{E}}(B_2) = 0.25$. She then learns evidence that imposes the constraint $P(R_1|R_1 \vee R_2) = 3/4$. Maximizing entropy then gives $P_{\mathcal{E}'}(R_1) = 0.351, P_{\mathcal{E}'}(R_2) = 0.117$, and $P_{\mathcal{E}'}(B_1) = P_{\mathcal{E}'}(B_2) = 0.266$. Grove and Halpern (1997) argue that by expanding the probability space Bayesian conditionalization can be applied to this problem; this update differs from the objective Bayesian update, though for slightly different reasons to the previous example (Uffink 1996, §6).[4] While the details will not be of concern to us here, it is important to note that this is also a case in which objective Bayesian updating and conditionalization disagree.

---

[4]Note that the Judy Benjamin problem was originally presented as a problem for minimum cross-entropy updating, yet another form of updating that incorporates aspects of both maximum entropy updating and Bayesian conditionalization (§3.4).

It has been claimed that in this example the maximum entropy update is counter-intuitive: the new evidence does not provide any reason to stop equivocating between red and blue, yet now $P_{\mathcal{E}'}(R_1 \vee R_2) = 0.468$ and $P_{\mathcal{E}'}(B_1 \vee B_2) = 0.532$. But this objection begs the question, since it assumes that one should continue to equivocate between red and blue if the evidence permits. Maximum entropy updating is not intended to be conservative: the idea is not to preserve features of prior degrees of belief when updating, but to calculate appropriate degrees of belief afresh. This requires being as equivocal as possible with respect to the space of elementary outcomes $\Omega = \{R_1, R_2, B_1, B_2\}$, not with respect to combinations of elementary outcomes such as $R_1 \vee R_2$ and $B_1 \vee B_2$. The new evidence is equivalent to the constraint $P(R_1) = 3P(R_2)$: the function satisfying this constraint that is maximally equivocal over Ω is simply not that which is maximally equivocal over red and blue. (Of course, if there were a good reason why one should continue to equivocate between red and blue, then this would yield a further constraint and the resulting probabilities would satisfy this constraint.) See Keynes (1921, chapter 4) for a discussion of why one should equivocate over the more refined partition.

4.2 Four kinds of incompatibility

In the literature thus far, any incompatibility between maximum entropy updating and Bayesian conditionalization has normally been taken to reflect negatively on the former rule rather than the latter (see, e.g. Friedman and Shimony 1971; Shimony 1973, 1986; Seidenfeld 1979, 1985; Dias and Shimony 1981; Skyrms 1985). Thus, the standard line of argument is: maxent is incompatible with Bayesian conditionalization, but conditionalization is well established and intuitive, and so maxent should be rejected.

This section will argue the opposite: that Bayesian conditionalization should be rejected because it disagrees with maximum entropy updating. The argument has two steps. First, we reformulate an old result which shows the conditions under which maxent and conditionalization agree. Next, we go through each of these conditions in turn and we see that where the condition fails, maxent is a more appropriate form of updating than conditionalization.

As before, suppose that the agent initially takes $\mathcal{E}$ for granted, that $\mathcal{E}$ imposes constraints χ, and that $P_\mathcal{E}$ is the probability function in $\mathbb{E} = \mathbb{P}_\chi$ that has maximum entropy. Then, the agent learns ε to give $\mathcal{E}' = \mathcal{E} \cup \{\varepsilon\}$ as her new evidence. This imposes the constraints χ' and $P_{\mathcal{E}'}$ is the new maximum entropy probability function.

Definition 4.1 ε *is* simple *with respect to* $\mathcal{E}$ *iff* $\chi' = \chi \cup \{P(\varepsilon) = 1\}$, *that is, the only constraint that ε imposes in the context of $\mathcal{E}$ is $P(\varepsilon) = 1$.*

The following is a reformulation of a result, versions of which are well known to both proponents and opponents of maximum entropy methods (Williams 1980; Seidenfeld 1986, Result 1):

Theorem 4.2 *If*

1. ε *is a sentence of the agent's language* $\mathcal{L}$,
2. ε *is simple with respect to* $\mathcal{E}$,
3. χ' *is consistent, and*
4. $P_\mathcal{E}(\cdot|\varepsilon)$ *satisfies* χ,

then $P_{\mathcal{E}'}(\theta) = P_\mathcal{E}(\theta|\varepsilon)$.

This result clearly states conditions under which the maximum entropy update will agree with the Bayesian conditionalization update of $P_\mathcal{E}^\mathcal{L}$. The following special case conveys the substantial agreement between the two forms of updating (here the initial evidence $\mathcal{E}$ is empty):

Corollary 4.3 *If the agent learns just consistent sentences $\varepsilon_1, \ldots, \varepsilon_n$ of $\mathcal{L}$ and each sentence ε_i is simple with respect to $\varepsilon_1, \ldots, \varepsilon_{i-1}$, then maximum entropy updating agrees with Bayesian conditionalization.*

See Seidenfeld (1986, Result 1 and corollary) for proofs of Theorem 4.2 and Corollary 4.3.

Incompatibilities between maxent and conditionalization can only arise if the conditions of Theorem 4.2 fail. Let us examine these conditions in turn.

4.2.1 Case 1

First, consider the case in which ε is not a sentence in the original language $\mathcal{L}$. In order to update, the language must of course be extended to include ε. Then, the objective Bayesian will update as normal by maximizing entropy on the new language $\mathcal{L}'$. However, the situation is not so straightforward for the subjectivist: it is not possible to apply Bayesian conditionalization because the prior $P_\mathcal{E}^{\mathcal{L}'}(\theta|\varepsilon)$ is not defined. The best one can do is to assign a 'prior' on the extended language post hoc, and then update by conditionalization. This is the strategy of the Olive–Sylvester and the Judy Benjamin examples discussed above. It is not a good strategy in general, because the new 'prior' is assigned at the same stage as posterior, so can be engineered to fit any desired posterior. In effect, conditionalization offers no constraint on the updating of probabilities if this post hoc prior strategy is pursued. To see this, suppose for example that the original language has two atomic states $\{R, \neg R\}$, where R signifies *the patient's cancer will recur*. An agent assigns prior probability $\frac{1}{2}$ to each proposition, then changes her degree of belief to 0.99 that the patient's cancer will recur, for no good reason at all. This is quite rational under the subjective Bayesian account if the above strategy is permitted. All the agent needs to do is find some proposition that is not expressible in the language but which is true (e.g. *the moon is not made of blue cheese*, or, *the first digit of π is 3*), extend her language to include sentences ε and $\neg \varepsilon$ expressing that proposition and its negation, and define a new post hoc 'prior' degree of belief $P(R|\varepsilon) = 0.99$; then her posterior degree of belief that the patient' cancer will recur is $P'(R) = 0.99$. The point of Bayesian conditionalization is to ensure that changes of degrees of belief are not so wild; clearly conditionalization achieves nothing in this case. Thus, this case favours maximum entropy updating over Bayesian conditionalization.

Note that if this post hoc prior strategy is implemented and then an incompatibility arises between maxent and conditionalization, this must be due to an infringement of one of the other conditions of Theorem 4.2, so we may move on to these other cases.

4.2.2 Case 2

Second, consider the case in which ε is not simple with respect to $\mathcal{E}$. An example will help show why maxent can disagree with conditionalization here. Suppose ε says '$P^*(\theta) = 0.8$', that is, that the chance of θ is 0.8. Then, by the Calibration norm, learning ε does not merely impose the constraint $P(\varepsilon) = 1$, but also the constraint $P(\theta) = 0.8$; hence, as long as $P(\theta) = 0.8$ is not already a constraint imposed by $\mathcal{E}$, ε is not simple with respect to $\mathcal{E}$. Should one prefer maxent or conditionalization in this situation? According to maxent, the agent will set $P_{\mathcal{E}'}(\theta) = 0.8$, which seems just right, given that the evidence imposes that constraint. On the other hand, if $P_\mathcal{E}(\theta|\varepsilon) \neq 0.8$, then it is quite wrong to conditionalize, because the new degree of belief will conflict with the new evidence. Once might try to save conditionalization by arguing that the agent *should* have set $P_\mathcal{E}(\theta|\varepsilon) = 0.8$ in the first place, and then no conflict would have arisen. But this argument fails because subjectivism permits the agent to set $P_{\mathcal{E}}(\theta|\varepsilon)$ to any value she likes—since ε is not granted at the time of setting the prior, it neither constrains the value of $P_{\mathcal{E}}(\theta)$ nor of $P_{\mathcal{E}}(\theta|\varepsilon)$. Now, if ε deductively entailed θ (respectively $\neg\theta$), then $P_{\mathcal{E}}(\theta|\varepsilon)$ would be constrained to 1 (respectively 0); but this is not the case here. (If the subjectivist would like to add a further norm that forces $P_{\mathcal{E}}(\theta|\varepsilon) = 0.8$, that is all well and good: no disagreement between the objective and subjective forms of updating will arise in this case, and we must focus on the other cases of disagreement. Of course, subjectivism will become that much less subjective as a consequence.)

Note that Bacchus et al. (1990, §4) argue that conditionalization goes wrong for just this reason; here is their example. Suppose A is 'Peterson is a Swede', B is 'Peterson is a Norwegian', C is 'Peterson is a Scandinavian', and ε is '80% of all Scandinavians are Swedes'. Initially, the agent sets $P_{\mathcal{E}}(A) = 0.2, P_{\mathcal{E}}(B) = 0.8, P_{\mathcal{E}}(C) = 1, P_{\mathcal{E}}(\varepsilon) = 0.2$, and $P_{\mathcal{E}}(A \wedge \varepsilon) = P_{\mathcal{E}}(B \wedge \varepsilon) = 0.1$. All these degrees of belief satisfy the norms of subjectivism. Updating by maxent on learning ε, the agent believes that Peterson is a Swede to degree 0.8, which seems quite right. On the other hand, updating by conditionalizing on ε leads to a degree of belief of 0.5 that Peterson is a Swede, which is quite wrong. Thus, we see that maxent is to be preferred to conditionalization in this kind of example because the conditionalization update does not satisfy the new constraints χ', while the maxent update does.

There is another way in which ε not being simple with respect to $\mathcal{E}$ can cause problems. In this scenario, both the conditionalization update and the maxent update satisfy χ', yet maxent is still to be preferred over conditionalization. Suppose A is 'Peterson is a Swede', C is 'Peterson is a Scandinavian', and ε is 'at least 50% of Scandinavians are Swedes and all politicians are liars'. $\mathcal{E}$ is the knowledge that C is true together with the testimony from a politician that $\neg A \wedge \varepsilon$ is false; so, $P_{\mathcal{E}}$ satisfies the constraints $\chi = \{P(C) = 1, P(\neg A \wedge \varepsilon) = 0\}$. Maximizing entropy, she sets $P_{\mathcal{E}}(A) = 2/3, P_{\mathcal{E}}(A|\varepsilon) = 1$. Then, the agent learns that ε is true. Now, ε is not simple with respect to $\mathcal{E}$—in particular, it undermines the politician's testimony. Plausibly, $\chi' = \{P(C) = 1, P(\varepsilon) = 1, P(A|\varepsilon) \geq 0.5\}$. Maximizing entropy now gives $P_{\mathcal{E}'}(A) = 0.5 \neq 1 = P_{\mathcal{E}}(A|\varepsilon)$. In this example, $P_{\mathcal{E}}(A|\varepsilon)$ does satisfy the constraints in χ', yet intuitively it is wrong to update by conditionalization because, given current evidence, to believe A to degree 1 is too extreme. Neither subjective nor objective Bayesians will want to *force* an agent to fully believe A when there is no evidence to warrant such a strong belief. If an updating rule is to be used to constrain the new degree of belief in A, then the maxent rule is much more plausible.

We see then that if ε is not simple with respect to $\mathcal{E}$, maxent is to be preferred over conditionalization where they differ, whether or not the conditionalization update satisfies the new constraints χ': if the conditionalization update does not satisfy constraints that should be satisfied, then it is clearly inappropriate; if it does satisfy the constraints, then either it agrees with the maxent update or it is less equivocal than the maxent update, but in the latter case it hardly seems right that an update rule should *force* one to have more extreme degrees of belief than the evidence warrants. Note that in the two examples considered earlier—Olive–Sylvester and Judy Benjamin—in which the language is extended to represent the new evidence, the new evidence is *not* simple with respect to the old evidence. The right thing to do in such a circumstance is to satisfy *all* the constraints imposed by the evidence (as maxent does), not just the constraint $P(\varepsilon) = 1$ (which is essentially all that Bayesian conditionalization does).

4.2.3 Case 3

The third kind of infringement of the conditions of Theorem 4.2 is the case in which χ' is inconsistent. This implies that either χ is inconsistent or ε is inconsistent with χ. In line with the restrictions imposed at the start of this chapter, we do not consider the former situation in any detail: here the consistency maintenance procedure will be carried out, $\mathbb{E} \neq \mathbb{P}_\chi$, and, as long as the other conditions hold, maxent will agree with conditionalization. The latter situation is the real problem area. If ε is inconsistent with χ, then $P_\mathcal{E}(\varepsilon) = 0$, and $P_\mathcal{E}(\theta|\varepsilon)$ is either undefined or unconstrained, according to how conditional probabilities are interpreted. (Recall that if, as in §2.3, conditional probability is *defined* by $P(\theta|\varepsilon) = P(\theta \wedge \varepsilon)/P(\varepsilon)$, then it is normally taken to be undefined when $P(\varepsilon) = 0$; alternatively if conditional probability is taken as primitive and subject to the constraint that $P(\theta|\varepsilon)P(\varepsilon) = P(\theta \wedge \varepsilon)$, then it is unconstrained when $P(\varepsilon) = 0$.) If $P_\mathcal{E}(\theta|\varepsilon)$ is undefined, then Bayesian conditionalization cannot be applied at all. If $P_\mathcal{E}(\theta|\varepsilon)$ is unconstrained, then the agent can assign it whatever value she likes, and Bayesian conditionalization offers no substantive constraint, in the sense that her new degree of belief in θ need in no way be connected to her unconditional prior degree of belief in θ, ε or indeed any other sentence. This leads to crazy changes in belief being deemed rational: for example, evidence that is known to count *against* θ can be invoked to *raise* the probability of θ. In sum, when ε is inconsistent with χ, conditionalization is inapplicable or vacuous. In contrast, maximum entropy updating is perfectly applicable in this situation: $P_{\mathcal{E}'}(\theta)$ is the probability function, from all those in $\mathbb{E}'$, that is most equivocal. Note that here $\mathcal{E}'$ is inconsistent so the consistency maintenance procedure will determine $\mathbb{E}'$ (which will of course differ from $\mathbb{P}_{\chi'} = \emptyset$). Hence, in this case the right way to update is via maxent and not via conditionalization.

4.2.4 Case 4

The fourth and final case is that in which $P_\mathcal{E}(\cdot|\varepsilon)$ does not satisfy χ. Here is an example from Seidenfeld (1986, §2). A die is thrown, the outcome space being $\Omega = \{1, \ldots, 6\}$. The agent knows that the expected score for this die is 3.5, so $\chi = \{\sum_\omega P(\omega) = 3.5\}$. Subject to just this constraint, $P_\mathcal{E}$ is uniform, giving $P_\mathcal{E}(\omega) = 1/6$ for each outcome. (The probability function from all those that satisfy the constraint which is closest to the equivocator is the equivocator itself, that is, the uniform distribution.) The agent then learns ε, that an even outcome occurred. If $\chi' = \{\sum_\omega P(\omega) = 3.5, P(\{2,4,6\}) = 1\}$, then we get

$P_{\mathcal{E}'}(2) = 0.46624$, $P_{\mathcal{E}'}(4) = 0.31752$, and $P_{\mathcal{E}'}(6) = 0.21624$. This differs from the result of Bayesian conditionalization since $P_{\mathcal{E}}(2|\varepsilon) = P_{\mathcal{E}}(4|\varepsilon) = P_{\mathcal{E}}(6|\varepsilon) = 1/3$. Note that the solution obtained by conditionalization does not satisfy the constraints χ, since the expectation is now 4, not 3.5. Which is right, maxent or conditionalization? If the analysis is right, and $\sum_\omega P(\omega) = 3.5$ really is a constraint on both the prior and the updated belief functions, then conditionalization *must* be wrong—it simply does not satisfy that constraint. On the other hand, suppose the analysis is wrong. If the constraint should not be in χ, then there need not be any disagreement between maxent and conditionalization after all—this will depend on whether $P_{\mathcal{E}}(\cdot|\varepsilon)$ satisfies the correct χ. Alternatively, if the constraint should be in χ but should not be in χ', then ε is not simple with respect to $\mathcal{E}$. This is the second case discussed above. There we saw that whether or not the conditionalization update satisfies the correct χ', it can hardly be considered the correct update where it differs from the maxent solution.

This concludes the argument. We have seen that in each of the four cases where maxent disagrees with conditionalization, the maxent update is to be preferred over the conditionalization update.

4.3 Criticisms of conditionalization

It is worth rehearsing some of the standard arguments for and against conditionalization in the light of the previous discussion. We take a look at two arguments *for* conditionalization in §§4.4, 4.5. In this section we will concentrate on arguments *against* conditionalization. The argument of §4.2 is only one of many that have been pitched against Bayesian conditionalization. It has been argued that conditionalization is problematic for the following reasons.

For one thing, conditionalization requires being eternally true to prior beliefs: an agent's degrees of belief throughout time are fixed by her initial probability distribution—in particular by conditional probabilities of the form $P(\theta|\varepsilon)$. Now, while the agent may deem these conditional degrees of belief appropriate at the outset, they may well turn out to be inappropriate in retrospect. Indeed, this is typically the situation when the new evidence is not simple with respect to prior knowledge (case 2 of §4.2). Thus, for example, exchangeable prior degrees of belief may be appropriate in the absence of knowledge about an experimental set-up, but when the string ε of evidence indicates that the data is being produced by a Markovian process, rather than a sequence of independent trials, exchangeable degrees of belief are a disaster (Gillies 2000a, pp. 77–83; Williamson 2007a); however, the agent is not permitted by conditionalization to change her degrees of belief to be non-exchangeable. (Recall from §2.2 that given a sequence of outcomes, positive or negative, an agent's prior degrees of belief are *exchangeable* iff for each number m, the agent assigns the same degree of belief to each possible sequence of outcomes that yields m positive outcomes in total.) Even advocates of conditionalization admit that this is unsatisfactory and that conditionalization may need to be overridden when a prior is re-evaluated (see, e.g. Lange 1999, §2; Earman 1992). But these advocates of conditionalization offer little in the way of a general rule for updating. Of course, a vague prescription like 'conditionalize unless an update that does not conform to conditionalization is warranted, in which case use the latter update' is hardly a rule at all. In light of the discussion of §4.2, it appears that the best way to formulate a precise rule of this form is to say 'conditionalize unless one of the conditions of Theorem 4.2 is infringed, in which case update by maximizing entropy'. But then it is equivalent and simpler just to say 'update by maximizing entropy'.

Note that maximum entropy updating—and objective Bayesian updating in general—does not suffer from this pathology: it does not force the agent to remain eternally true to her prior beliefs. At each step, the agent chooses a belief function, from all those that fit the evidence at the time, that is maximally equivocal. There is simply no room for the agent's updated degrees of belief to be out of step with the evidence. This fact is often overlooked: it is often charged that objective Bayesianism suffers from being stuck with an unreasonable prior, namely one that renders learning from experience impossible (see, e.g. Dias and Shimony 1981, §4). Consider elementary propositions $B_1, \ldots, B_{101}$, where B_i signifies that the ith observed raven is black. Subject to no constraints, maximizing entropy yields $P_\emptyset(B_{101}) = P_\emptyset(B_{101}|B_1 \wedge \cdots \wedge B_{100}) = 1/2$. So it appears that observing 100 black ravens does not strengthen an agent's belief that raven 101 is black—that is, the agent fails to learn from experience. But this objection is misplaced, since $P_\emptyset(B_{101}|B_1 \wedge \cdots \wedge B_{100})$ does *not* represent the degree to which the agent should believe that raven 101 is black on the basis of evidence $B_1 \wedge \cdots \wedge B_{100}$. This is because the evidence $B_1 \wedge \cdots \wedge B_{100}$ is not simple. Not only does learning this evidence impose the constraint $P(B_1 \wedge \cdots \wedge B_{100}) = 1$, but also the constraint that $P(B_{101}) \simeq 1$. (Exactly how this last constraint is to be made precise is a question of statistical inference—the details need not worry us here.) When evidence $B_1 \wedge \cdots \wedge B_{100}$ constitutes the new evidence $\mathcal{E}'$, we see that $P_\emptyset(B_{101}) = 1/2$ (and $P_\emptyset(B_{101}|B_1 \wedge \cdots \wedge B_{100}) = 1/2$), but $P_{\mathcal{E}'}(B_{101}) \simeq 1$. Hence, this is a case in which maximum entropy updating disagrees with conditionalization, and the agent does, after all, learn from experience.

The above objection to conditionalization is often put another way: conditionalization requires specifying all future degrees of belief in advance—that is, specifying from the outset $P(\theta|\varepsilon)$, for all possible sequences ε of evidence—but this is impossible from a practical point of view. Note that the Maximum Entropy Principle does not require that so many probabilities be specified. Since degrees of belief are a function of the agent's evidence $\mathcal{E}$, she does not need to calculate the full function $P_\mathcal{E}$ at any particular time, but only that part of the function that is required for practical purposes. Thus, at any stage $P_\mathcal{E}$ may be partially specified—because it is determined afresh on changes to $\mathcal{E}$, this does not prevent updating. The subjectivist may appeal by saying that she can also partially specify her belief function, and just specify the 'prior' $P(\theta|\varepsilon)$ after the evidence ε has been discovered. But our discussion in §4.2 of the post-hoc-prior strategy shows that this does not save conditionalization. Moreover, if ε is not simple with respect to prior knowledge, then for a post hoc 'prior' $P(\theta|\varepsilon)$ to be reasonable, it can typically not really be construed as part of the prior at all—the reasonable value of $P(\theta|\varepsilon)$ will in general not be the value that would have been chosen prior to the gathering of evidence—it would be the maxent update instead. Contrapositively, if the post hoc $P(\theta|\varepsilon)$ were to match the genuine prior value, then either it would agree with maxent or it would not lead to a reasonable update.

Closely related to this objection is the charge that Bayesian conditionalization requires every possible future piece of evidence to be representable as a sentence in the agent's prior language. It is clearly unreasonable to demand that an agent have this amount of foresight. As we saw, the natural response for the subjectivist is to claim that one can extend the language to incorporate the new evidence as and when it becomes known, and then to give a 'prior' value to $P(\theta|\varepsilon)$ post hoc. But, to reiterate, this is a poor strategy because it offers no constraint on the post hoc degrees of belief. As we also saw, maxent does not succumb to this problem. If the new evidence is not in the old language then the language can simply be extended to incorporate ε and maxent performed as normal (Williamson 2005a, chapter 12). Since on the objective Bayesian account posterior degrees of belief are constrained by maxent rather than by prior degrees of belief, there is no need for evidence to be representable in the prior language.

Bayesian conditionalization has also been criticized on account of the fact that conditionalization must render all evidence eternally certain, even though evidence is clearly fallible. When ε is learnt, $P(\varepsilon)$ remains 1 for all time, even if later evidence casts doubt on ε. There does not seem to be any way for the subjectivist to uphold Bayesian conditionalization yet get round this objection. It is sometimes argued that ε should not have been considered conditionalizable evidence if it were not beyond all possible doubt. (If say $P(\varepsilon) = 0.9$, then Jeffrey conditionalization, another special case of maxent updating, could be applied instead of Bayesian conditionalization.) But of course this response is tantamount to abandoning Bayesian conditionalization, since practically no evidence is incontrovertible. On the other hand, maxent does not render evidence eternally certain: if the veracity of ε comes into doubt, one simply removes ε from the current evidence, and then maximizes entropy as usual. (Indeed, while in this book we maintain that all evidence is certain in the agent's current operating context, from a formal point of view maxent itself does not require that a given input ε be certain. If ε is obtained by a process with error rate δ, then learning ε imposes the constraint $P(\varepsilon) = 1 - \delta$, rather than $P(\varepsilon) = 1$. Maxent can handle such constraints with no difficulty, while Bayesian conditionalization requires that $P(\varepsilon) = 1$ when ε is learned.)

Colin Howson has argued compellingly that Bayesian conditionalization is inconsistent in some situations (Howson 1997, §3, 2000, p. 136). Consider the following example, cast in subjectivist terms. Today you are certain of θ, so $P(\theta) = 1$, but suppose you think that tomorrow you may be significantly less than certain about θ, that is, $P(P'(\theta) \leq 0.9) > 0$. Since $P(\theta) = 1, P(\theta|P'(\theta) \leq 0.9) = 1$. Then, tomorrow comes and indeed $P'(\theta) \leq 0.9$, and you realize this by introspection. But by conditionalization, $P'(\theta) = P(\theta|P'(\theta) \leq 0.9) = 1$, a contradiction. Note that here the evidence $P'(\theta) \leq 0.9$ is not simple, since it imposes the constraint $P'(\theta) \leq 0.9$ as well as the constraint $P'(P'(\theta) \leq 0.9) = 1$. So in the light of our preceding discussion it is no wonder that conditionalization goes wrong. An objective Bayesian approach is clearly to be preferred here: if $P(\theta) \leq 0.9$ is genuinely a constraint on new degrees of belief, then it will be satisfied with no contradiction; if it is not genuinely a constraint, then it will not be imposed and again no contradiction will arise.

4.4 A Dutch book for conditionalization?

Having set the case against conditionalization, it is only fair to briefly consider a couple of the more widely cited arguments in favour of conditionalization: the argument from conservativity (discussed in §4.5) and the diachronic Dutch book argument for conditionalization (discussed in this section).

Paul Teller presents the Dutch book argument in Teller (1973) attributing the idea to David Lewis. The argument aims to show that if you fail to update by conditionalization then you are susceptible to a Dutch book, that is, if you bet according to your degrees of belief, then a judicious bookmaker can make you lose money whatever happens. This appears to be a clear reason to favour conditionalization over maxent. Van Fraassen (1989, p. 174) points out that this Dutch book argument only shows that *if you specify an updating rule in advance* and that rule is not conditionalization, then you are susceptible to Dutch book; but this is exactly the case we have here: maxent is a rule, specified in advance, that may disagree with conditionalization. Thus, this Dutch book argument appears to be a serious challenge to maxent.

We will not go through the details of the Dutch book argument here, since no doubt will be cast on its validity. Instead, doubt will be cast on the importance of diachronic Dutch books in general, by means of a reductio. Specifically, we shall see that in certain situations one can Dutch book *anyone who changes their degrees of belief at all*, regardless of whether or not they change them using conditionalization. Thus, avoidance of Dutch book is a lousy criterion for deciding on an update rule.[5]

Suppose it is generally known that you will be presented with evidence that does not count against θ, so that your degree of belief in θ will not decrease. It is also known that the evidence is inconclusive, so that you will not be certain about θ. It is not known in advance what this evidence is. (This is the kind of situation which might arise in a court of law, for example, when a juror hears

---

[5]It should be emphasized that this is only an attack on dynamic Dutch books, that is, Dutch books involving degrees of belief at more than one time, not on synchronic Dutch books such as the Dutch book argument for the axioms of probability, which remain compelling. There is a large literature concerning the cogency of Dutch book arguments. See, for example, Howson (1997, §7) for related worries about dynamic Dutch books, Foley (1993, §4.4) for reservations about Dutch book arguments in general, Rowbottom (2007) for a recent critique of the synchronic Dutch book argument in the subjectivist framework, and Skyrms (1987) for a defence of Dutch book arguments in general.

that the prosecution will present its evidence, knowing that the prosecution is competent enough not to present evidence that counts against its case.) You set your prior $P(\theta) = q$. Now, in the betting set-up that we encountered in §§3.1, 3.2, this is tantamount to a bet which will yield S if θ occurs for a payment of qS, where S is a real-valued stake chosen by a bookmaker that may be positive or negative. Thus, your loss is $L = (q - I_\theta)S$, where I_θ is the indicator function for θ. The stake-maker chooses $S = -1$. Then, the new evidence arrives and you duly set $P'(\theta) = q'$, where $q \leq q' < 1$. This is interpreted as a second bet, with stake S'. Your total loss over the prior and posterior bets on θ is $L' = (I_\theta - q) + (q - I_\theta)S'$. Now, unless $q' = q$, the stake-maker can choose stake S' such that $q/q' < S' < (1-q)/(1-q')$ to ensure that $L' > 0$ whatever happens. Thus, unless your degree of belief in θ remains constant, you are susceptible to a Dutch book.

In conclusion, do not trust dynamic Dutch books: it can be more important to change degrees of belief to reflect new evidence than to avoid certain loss.

4.4.1 Conditionalization and conditional belief

One might object to the above reductio along the following lines.[6] The knowledge that you will be presented with evidence that does not count against θ might be cashed out as the claim that there is a partition Π of propositions such that you will be presented with one of these propositions as evidence and $P(\theta|\varepsilon) \geq P(\theta)$, for each $\varepsilon \in \Pi$. Now, if there were one or more $\varepsilon \in \Pi$ such that $P(\theta|\varepsilon) > P(\theta)$, then by the axioms of probability,

$$P(\theta) = \sum_{\varepsilon \in \Pi} P(\theta|\varepsilon) P(\varepsilon) > \sum_{\varepsilon \in \Pi} P(\theta) P(\varepsilon) = P(\theta),$$

assuming the conditional probabilities are well defined, that is, $P(\varepsilon) > 0$, for all $\varepsilon \in \Pi$. But this is of course a contradiction. Hence, it must be the case that $P(\theta|\varepsilon) = P(\theta)$, for all $\varepsilon \in \Pi$. So it is the laws of probability rather than dynamic Dutch book considerations that rule out changes in degree of belief: the reductio is redundant.

This line of argument is unsuccessful because the knowledge that you will be presented with evidence that does not count against θ is more correctly cashed out as: for ε^*, the evidence you will in fact be presented with next, $P'(\theta) \geq P(\theta)$, where P' is the belief function you will have after learning ε^*.

In particular, one cannot presume that $P'(\theta) = P(\theta|\varepsilon^*)$ since that is just the conditionalization principle, which is exactly what is under question here.

Moreover, there is no obvious partition Π for which $P(\theta|\varepsilon) \geq P(\theta)$, for each $\varepsilon \in \Pi$. One might try to choose $\Pi = \{\varepsilon^*, \neg\varepsilon^*\}$ in which case

$$P(\theta) = P(\theta|\varepsilon^*)P(\varepsilon^*) + P(\theta|\neg\varepsilon^*)P(\neg\varepsilon^*).$$

---

[6]I am very grateful to Corey Yanofsky for suggesting this line of argument, and to Bas van Fraassen for a similar suggestion (van Fraassen 2009, §2).

But then one can only conclude that if $P(\theta|\varepsilon^*) > P(\theta)$, then $P(\theta|\neg\varepsilon^*) < P(\theta)$, which is by no means at odds with the claims of the reductio.

But suppose there is some appropriate partition Π such that $P(\theta|\varepsilon) \geq P(\theta)$, for each $\varepsilon \in \Pi$—what can we conclude then? We can surely conclude that $P(\theta|\varepsilon) = P(\theta)$, for all $\varepsilon \in \Pi$, but what significance does this have?

If we are careful to avoid presuming conditionalization itself, the conclusion that $P(\theta|\varepsilon) = P(\theta)$, for all $\varepsilon \in \Pi$, fails to say that your degree of belief in θ should not change, which is the conclusion of the reductio. So the reductio is not, after all, redundant.

Furthermore, if one does not presume conditionalization, then conditional probabilities are not even naturally interpretable as conditional beliefs. If one commits to conditionalization, then one commits to the view that $P(\theta|\varepsilon)$ represents your degree of belief in θ should you grant that ε is true. If one rejects conditionalization as a universal principle, then one rejects this interpretation, and $P(\theta|\varepsilon)$ is merely an abbreviation of $P(\theta \wedge \varepsilon)/P(\varepsilon)$. In particular, $P(\theta) = P(\theta|\varepsilon)$ is not to be understood as the claim that your degree of belief in θ is identical to your degree of belief in θ should ε be granted, but just as the claim that $P(\theta)P(\varepsilon) = P(\theta \wedge \varepsilon)$ and $P(\varepsilon) \neq 0$, that is, just as the formal claim that current degrees of belief in θ and ε are probabilistically independent. (Note that when a conditional probability is interpreted as a conditional belief, the conditional belief may in turn be interpreted as a *conditional bet*: $P(\theta|\varepsilon)$ is then interpreted as a betting quotient on θ with the proviso that the bet is called off if ε turns out to be false. Our rejection of the conditional belief interpretation of conditional probabilities is thus accompanied by a rejection of the conditional bet interpretation. Conditional probabilities act as abbreviations, nothing more.)

In sum, as long as one is careful not to beg the question by presuming either the conditionalization principle itself or the associated claim that conditional probabilities quantify conditional beliefs, the above reductio remains unscathed.

4.4.2 A note on reflection

A rather different Dutch book argument to the above-mentioned reductio was put forward by van Fraassen (1984) in favour of the *Principle of Reflection*, which holds that $P(\theta|P'(\theta) = r) = r$. That argument concerns how one should set one's prior degrees of belief, while the argument presented above concerns posterior degrees of belief. There is another difference between the two arguments: here we do not assume that statements about her future beliefs are expressible in the agent's language, so we do not assume that they are in the domain of the agent's belief function. If the reductio outlined above is accepted, van Fraassen's diachronic Dutch book argument for Reflection should be treated with suspicion.

More than that though, the Principle of Reflection itself should be treated with caution. Consider the following extension of the above betting scenario. Suppose first that the agent can express propositions about her future beliefs, so that it makes sense to consider expressions like $P(P'(\theta) = r)$. Suppose too that she knows that her degree of belief in θ should either remain the same or should increase to some value r. Now by the Principle of Reflection,

$$q = P(\theta)$$
$$= P(\theta|P'(\theta) = q)P(P'(\theta) = q) + P(\theta|P'(\theta) = r)P(P'(\theta) = r)$$
$$= qx + r(1-x),$$

where $x = P(P'(\theta) = q)$. But for this identity to hold either $x = 1$ or $r = q$. Either case implies that the agent should fully believe that her degree of belief in θ will not change. But that seems a ridiculously strong conclusion. Is this a reductio of the Principle of Reflection?

The proponent of the Principle of Reflection might suggest that all this shows is that the agent's current degree of belief should not be q; perhaps, it should be somewhere between q and r, the possible values of her future degree of belief in θ. But in the example you know that your degree of belief should either remain the same or increase to a fixed r. The proponent's suggestion rules out the former possibility. Hence, the agent's degree of belief should increase to r and, by Reflection, her present degree of belief should be r. Far from escaping the reductio, the proponent of Reflection has ended up in a rather worse situation. By trying to avoid the conclusion that the agent should believe that her degree of belief in θ should not change, we have reached the conclusion that her degree of belief in θ should be greater than is independently plausible *and* should not change.

In fact, the Principle of Reflection and Bayesian conditionalization come unstuck for similar reasons. Reflection is intended to capture the following idea: if one were to learn that one will come to believe θ to degree r, then one should currently believe θ to degree r. But $P(\theta|P'(\theta) = r) = r$ fails to express this idea. The conditioning proposition $P'(\theta) = r$ is not simple; hence this is a case where conditionalization can break down and the degree to which one ought to believe θ, were one to grant that $P'(\theta) = r$, should not be expected to equal $P(\theta|P'(\theta) = r)$. This analysis motivates reformulating the Reflection principle to more adequately express the above idea; perhaps as $P_{\mathcal{E} \cup \{P'(\theta)=r\}}(\theta) = r$. From this new Principle of Reflection one can no longer derive the above reductio.

4.5 Conditionalization from conservativity?

Another justification of Bayesian conditionalization proceeds along the following lines. By the axioms of probability,

$$P'(\theta) = P'(\theta|\varepsilon)P'(\varepsilon) + P'(\theta|\neg\varepsilon)P'(\neg\varepsilon) = P'(\theta|\varepsilon) \text{ if } P'(\varepsilon) = 1.$$

So conditionalization is the only possible way to update if $P'(\theta|\varepsilon) = P(\theta|\varepsilon)$. Now, plausibly, any agent should be conservative in the sense that she should keep $P'(\theta|\varepsilon) = P(\theta|\varepsilon)$ if at all possible. But it is always possible to be conservative in this sense if all one learns is ε. Hence, conditionalization is the right way to update.

This argument fails for the following reason: the kind of conservativity it appeals to is far too strong. Plausibly, perhaps, any agent should be conservative in the sense that she should keep $P'(\theta|\varepsilon) = P(\theta|\varepsilon)$ *unless the evidence ε indicates otherwise*. But the evidence may indicate otherwise—the evidence may not be simple with respect to the previous evidence (Case 2 of §4.2), or $P'(\theta) = P(\theta|\varepsilon)$ may not satisfy the original constraints χ (Case 4). Of course, in the case in which ε is not in the event space then the above argument does not succeed either (Case 1). Nor does it work if $P(\varepsilon) = 0$ (Case 3). The best one can say then is that updating should agree with conditionalization if the conditions of Theorem 4.2 hold.

In sum, when the conservativity condition is rendered less dubious, this justification fails to tell us anything we did not already know on the basis of the discussion of §4.2.

4.6 Summary

We saw in §4.2 that if we take the trouble to look at the different scenarios in which maxent disagrees with conditionalization, maxent is to be preferred over conditionalization in each case. Hence, the argument against maximum entropy updating on the grounds of the existence of such disagreements backfires on the subjectivist.

This analysis helps to shed some light on some of the standard complaints against conditionalization (§4.3). While these remain cogent criticisms of conditionalization, maxent does not succumb to any of these objections. On the other hand, two key justifications of conditionalization turn out to be less than cogent. The Dutch book justification of conditionalization fails because diachronic Dutch books are subject to a reductio (§4.4). The argument from conservativity fails to offer a fully general justification of conditionalization—it only applies where conditionalization agrees with maxent (§4.5). Thus, the case against conditionalization is stronger than ever.

But without a viable form of updating, subjectivism is laissez-faire to the point of ridicule: not only is an agent rationally entitled to adopt practically any probability function as her prior, she is rational to change it in any way she wishes too. In particular, under a subjectivist account without updating if a consultant has no evidence which bears on whether a patient's cancer will recur or not then she is not only rational to initially believe that the patient's cancer will recur to degree 1 but also rational to change her degree of belief to 0 on a whim.

5

Predicate Languages

Thus far we have considered objective Bayesian probability to be defined over finite propositional languages. In this chapter, we extend the formalism to predicate languages. As well as reaping the rewards of being able to handle richer languages, this extension also allows us to invoke objective Bayesian epistemology as a semantics for probabilistic logic in Chapter 7.

The required extension to predicate languages is summarized in §5.1. Subsequent sections of this chapter are devoted to motivating and discussing this new formalism.

5.1 The framework

We suppose now that $\mathcal{L}$ is a first-order predicate language without equality. It is convenient to assume that each individual is picked out by a unique constant symbol t_i; we suppose that there is a countable infinity $t_1, t_2, \ldots$ of such constants, but only finitely many predicate symbols. For $n \geq 1$, let $\mathcal{L}_n$ be the finite predicate language involving the symbols of $\mathcal{L}$ but with only finitely many constants $t_1, \ldots, t_n$. Let $A_1, A_2, \ldots$ run through the atomic propositions of $\mathcal{L}$, that is, propositions of the form Ut, where U is a predicate symbol and t is a tuple of constant symbols of the corresponding arity. We insist that the $A_1, A_2, \ldots$ are ordered as follows: any atomic proposition expressible in $\mathcal{L}_n$ but not expressible in $\mathcal{L}_m$ for $m < n$ should occur later in the ordering than those atomic propositions expressible in $\mathcal{L}_m$. Let $A_1, \ldots, A_{r_n}$ be the atomic propositions expressible in $\mathcal{L}_n$. An *atomic n-state* ω_n is an atomic state $\pm A_1 \wedge \cdots \wedge \pm A_{r_n}$ of $\mathcal{L}_n$. Let Ω_n be the set of atomic n-states.

As before, we base objective Bayesian epistemology on the norms of Probability, Calibration, and Equivocation.

Probability: The strengths of the agent's beliefs should be representable by a probability function.

This norm now requires that there be a probability function P on predicate language $\mathcal{L}$ such that $P(\theta)$ represents the degree to which the agent should believe θ, for each sentence θ of $\mathcal{L}$. Here a *probability function* on a predicate language $\mathcal{L}$ is a function $P : S\mathcal{L} \longrightarrow \mathbb{R}$ that satisfies the properties

PP1: $P(\omega_n) \geq 0$, for each $\omega_n \in \Omega_n$ and each n,

PP2: $P(\tau) = 1$, for some tautology $\tau \in S\mathcal{L}$,

PP3: $P(\theta) = \sum_{\omega_n \models \theta} P(\omega_n)$, for each quantifier-free proposition θ, where n is large enough that $\mathcal{L}_n$ contains all the atomic propositions occurring in θ, and

PP4: $P(\exists x \theta(x)) = \sup_m P\left(\bigvee_{i=1}^m \theta(t_i)\right)$

Note, in particular, that a probability function P on predicate language $\mathcal{L}$ is determined by its values on the ω_n, for $n = 1, 2, \ldots$ (see, e.g. Paris 1994, theorem 11.2). PP4 is known as *Gaifman's condition*, and PP1–4 imply that $P(\exists x \theta(x)) = \lim_{m \to \infty} P\left(\bigvee_{i=1}^m \theta(t_i)\right)$ and $P(\forall x \theta(x)) = \lim_{m \to \infty} P\left(\bigwedge_{i=1}^m \theta(t_i)\right)$.

Calibration: The agent's degrees of belief should satisfy constraints imposed by her evidence.

Again it is supposed that there is some set $\mathbb{E}$ of probability functions on $\mathcal{L}$ that are compatible with evidence $\mathcal{E}$ and that the probability function $P_\mathcal{E}$ representing the agent's degrees of belief should lie in that set. As before, we take $\mathbb{E} = \langle \mathbb{P}^* \rangle \cap \mathbb{S}$: for a belief function to be compatible with evidence it should lie in the convex hull of the chance functions compatible with evidence and should satisfy structural constraints imposed by qualitative evidence. Thus, the Calibration norm carries straight over from the propositional case.

Equivocation: The agent's degrees of belief should otherwise be sufficiently equivocal.

In the case of a predicate language, we can define the equivocator $P_=$ on $\mathcal{L}$ by

$$P_=(\omega_n) = \frac{1}{2^{r_n}},$$

for all ω_n. Clearly, this function is the only function that equivocates between the atomic n-states, for each fixed n. We consider the n-*divergence* between probability functions,

$$d_n(P, Q) = \sum_{\omega_n \in \Omega_n} P(\omega_n) \log \frac{P(\omega_n)}{Q(\omega_n)},$$

where as before $0 \log 0$ is taken to be 0. We say that P is *closer* to R than Q if there is some N such that for all $n \geq N$, $d_n(P, R) < d_n(Q, R)$. We write $P \prec Q$ if P is closer to the equivocator $P_=$ than Q. Now, we define $\downarrow\mathbb{E}$ to be the set of members of $\mathbb{E}$ that are minimal with respect to $\prec$, $\downarrow\mathbb{E} \stackrel{df}{=} \{P \in \mathbb{E} : \text{there is no } Q \in \mathbb{E} \text{ such that } Q \prec P\}$. Objective Bayesianism then requires that the agent's degrees of belief be representable by a probability function $P_\mathcal{E} \in \Downarrow\mathbb{E}$, that is, a probability function sufficiently close to $P_=$. As before, $\downarrow\mathbb{E} \subseteq \Downarrow\mathbb{E}$; these two sets may be expected to differ for pragmatic reasons, and will certainly differ in cases where $\downarrow\mathbb{E}$ is empty.

Define the n-*entropy* $H_n(P)$ by

$$H_n(P) = -\sum_{\omega_n \in \Omega_n} P(\omega_n) \log P(\omega_n).$$

We say that P has *greater entropy* than Q, written $P \gg Q$, if there is some N such that for all $n \geq N$, $H_n(P) > H_n(Q)$. The claim that $P_\mathcal{E} \in \downarrow\mathbb{E}$ is equivalent to:

92   *In Defence of Objective Bayesianism*

Maximum Entropy Principle: The agent's degrees of belief should be representable by a probability function $P_\mathcal{E} \in \{P \in \mathbb{E} : P \text{ is maximal with respect to} \gg\}$ in cases where there is such a maximizer.

But in general we merely require that the entropy of $P_\mathcal{E}$ be *sufficiently* great, where pragmatic considerations are used to determine exactly what counts as sufficiently great. And, as discussed in §3.4, we note that these pragmatic considerations might appeal to notions of distance (and hence entropy) that are different to those defined above, though we focus on cross-entropy divergence in this book.

Having presented the principal definitions, in the remainder of this chapter we shall discuss their motivation and the key properties of the framework.

5.2   The Probability norm

The new version of the Probability norm ascribes probabilities to existentially and universally quantified propositions and, as discussed in §3.2, these probabilities are interpreted via betting considerations. One might think that there is a tension here, because a bet on a quantified proposition may never be settled. If there are infinitely many elements of the domain, then normally one cannot tell in a finite time that a universally quantified proposition is true or that an existentially quantified proposition is false. (There are exceptions, though—for instance, if the proposition is a tautology or a contradiction.) In which case, a bet on the truth of such a proposition will not be settled and so betting considerations can hardly motivate a particular value for the agent's degree of belief in the proposition. Hence, it appears that one can assign such a proposition any degree of belief at all, contrary to the constraints imposed by PP1–4.

There are two possible responses to this worry. First, one might point out that PP4 can be motivated by semantics instead of betting. The identity invoked by PP4 should hold in virtue of the meaning of 'there exists', given our assumption that each member of the domain is picked out by some constant symbol (Paris 1994, pp. 162–3). This identity need not be justified by a Dutch book argument, and so its justification does not require that bets on quantified propositions be settled in principle.

A second possible response involves denying the tension. The betting scenario is in any case a considerable idealization of the following form: 'assuming an agent for whom the utility of money increases linearly with its size were to bet for unknown positive or negative stakes, and assuming there were a stake-maker with exactly the same evidence as the agent trying to force the agent to lose money, and assuming that the agent cares equally about losses of the same magnitude whenever they are incurred in the future, and assuming the settlor of the bets makes no mistakes...'. It does not take much to add the condition 'assuming that the settlor of the bets is omniscient'. In which case, one could, if one wished, justify PP4 by Dutch book considerations after all:

Theorem 5.1 *Define function $P : S\mathcal{L} \longrightarrow \mathbb{R}$ by $P(\theta) = $ the agent's betting quotient for θ. The agent's bets on propositions expressible in $\mathcal{L}$ avoid the possibility of a Dutch book iff P is a probability function,*

PP1: $P(\omega_n) \geq 0$, for each $\omega_n \in \Omega_n$ and each n.
PP2: $P(\tau) = 1$, for some tautology $\tau \in S\mathcal{L}$,
PP3: $P(\theta) = \sum_{\omega_n \models \theta} P(\omega_n)$, for each quantifier-free proposition θ, where n is large enough that $\mathcal{L}_n$ contains all the atomic propositions occurring in θ
PP4: $P(\exists x \theta(x)) = \sup_m P(\bigvee_{i=1}^m \theta(t_i))$.

Proof: First, we show that if the agent's betting quotients are not Dutch-bookable, then they satisfy the axioms of Probability.

That PP1–3 hold follows just as in Theorem 3.2.

PP4. Suppose that $P(\exists x \theta(x)) \neq \sup_m P(\bigvee_{i=1}^m \theta(t_i))$; we show that the agent is susceptible to a Dutch book. If $P(\exists x \theta(x)) < \sup_m P(\bigvee_{i=1}^m \theta(t_i))$, then $P(\exists x \theta(x)) < P\left(\bigvee_{i=1}^M \theta(t_i)\right)$, for some M. Set stakes

$$S_\varphi = \begin{cases} -1 & : \quad \varphi = \exists x \theta(x) \\ 1 & : \quad \varphi = \theta(t_1) \vee \cdots \vee \theta(t_M) \\ 0 & : \quad \text{otherwise} \end{cases}.$$

The agent's total loss is then

$$L = -\left[P(\exists x \theta(x)) - I_{\exists x \theta(x)}\right] + \left(P(\theta(t_1) \vee \cdots \vee \theta(t_M)) - I_{\theta(t_1) \vee \cdots \vee \theta(t_M)}\right).$$

But this is positive whatever happens because $I_{\exists x \theta(x)} \geq I_{\theta(t_1) \vee \cdots \vee \theta(t_M)}$, so we have a Dutch book.

On the other hand, if $P(\exists x \theta(x)) > \sup_m P(\bigvee_{i=1}^m \theta(t_i))$, then set

$$S_\varphi = \begin{cases} 1 & : \quad \varphi = \exists x \theta(x) \\ (-1)^k & : \quad \varphi = \theta(t_{i_1}) \wedge \cdots \wedge \theta(t_{i_k}) \\ 0 & : \quad \text{otherwise} \end{cases}.$$

Now, by the addition rule of probability (which follows from PP1–3) we have that

$$P\left(\bigvee_{i=1}^m \theta(t_i)\right) = \sum_{\{i_1,\ldots,i_k\} \subseteq \{1,\ldots,m\}} (-1)^{k+1} P\left(\theta(t_{i_1}) \wedge \cdots \wedge \theta(t_{i_k})\right).$$

Hence,

$$\sup_m P\left(\bigvee_{i=1}^m \theta(t_i)\right) = \sum_{\{i_1,\ldots,i_k\} \subseteq \mathbb{N}} (-1)^{k+1} P\left(\theta(t_{i_1}) \wedge \cdots \wedge \theta(t_{i_k})\right).$$

Moreover,

$$I_{\exists x\theta(x)} = \sum_{\{i_1,\ldots,i_k\}\subseteq \mathbb{N}} (-1)^{k+1} I_{\theta(t_{i_1})\wedge\cdots\wedge\theta(t_{i_k})}.$$

So the agent's total loss is

$$L = P(\exists x\theta(x)) - I_{\exists x\theta(x)} - \left[\sup_m P\left(\bigvee_{i=1}^m \theta(t_i)\right) - I_{\exists x\theta(x)}\right] > 0$$

whatever happens, as required.

Conversely, if the agent's betting quotients at any one time satisfy the axioms of probability, then they are not Dutch-bookable, for the following reason. We saw above that if PP4 holds, then the agent's bet on $\exists x\theta(x)$ is equivalent to a combination of her bets on the conjunctions $\theta(t_{i_1}) \wedge \cdots \wedge \theta(t_{i_k})$; more generally, her bet on a sentence θ that contains one or more quantifiers is equivalent to a combination of her bets on quantifier-free sentences. Hence, if PP4 holds, then betting on all of $S\mathcal{L}$ is tantamount to betting on quantifier-free sentences but double-counting some bets. But if PP1–3 hold, then bets on quantifier-free sentences are immune from Dutch book—this is demonstrated in the proof of Theorem 3.2 and this demonstration extends straightforwardly to the case in which double-counting (i.e. multiple bets on the same sentence with the same betting quotient) is permitted.

To spell this last point out, suppose the agent bets k_θ times on each quantifier-free sentence θ in her language using betting quotients that satisfy the axioms of probability. Should n-state ω_i turn out true, the agent's loss on those sentences expressible in $\mathcal{L}_n$ is

$$L_i^n = \sum_{\omega_i \models \theta \in S\mathcal{L}_n} k_\theta(P(\theta)-1)S_\theta + \sum_{\omega_i \not\models \theta \in S\mathcal{L}_n} k_\theta P(\theta) S_\theta$$
$$= \sum_{\theta \in S\mathcal{L}_n} k_\theta P(\theta) S_\theta - \sum_{\omega_i \models \theta \in S\mathcal{L}_n} k_\theta S_\theta$$

and her total loss is $L = \lim_{n\to\infty} L_i^n$. (Recall that we restrict attention to bets in which only a finite amount of money can change hands, so the above sums and limit are all defined and finite.)

Now, if this loss were positive in all eventualities, then the limiting expected loss,

$$\lim_{n\to\infty} \sum_{\omega_i \in \Omega_n} P(\omega_i) L_i^n,$$

would have to be positive. But we know from the proof of Theorem 3.2 that $\sum_{\omega_i \in \Omega_n} P(\omega_i) L_i^n = 0$, so the limiting expected loss is in fact zero. Hence, there is no possible Dutch book. ∎

5.3 Properties of the closer relation

Next, we investigate the *closer* relation defined in §5.1, showing that this notion does what one would expect of a closeness relation.

Proposition 5.2 *For fixed R the binary relation $\cdot$ is closer than $\cdot$ to R is irreflexive, asymmetric, and transitive.*

Proof: Irreflexivity is immediate from the definition: $d_n(P,R) \not< d_n(P,R)$.

Asymmetry is also immediate: if $d_n(P,R) < d_n(Q,R)$, then $d_n(Q,R) \not< d_n(P,R)$. For transitivity, suppose P is closer than Q to R and Q is closer than S to R. Then, there is some L such that for $n \geq L$, $d_n(P,R) < d_n(Q,R)$, and there is some M such that $n \geq M$ implies $d_n(Q,R) < d_n(S,R)$. Take N to be the maximum of L and M. Then, for $n \geq N$, $d_n(P,R) < d_n(S,R)$, so P is closer than S to R. ∎

Theorem 5.3 *If $P \neq Q$ and if, for sufficiently large n, $d_n(P,R) \leq d_n(Q,R)$, then any proper convex combination of P and Q, that is, $S = \lambda P + (1-\lambda)Q$, for $\lambda \in (0,1)$, is closer than Q to R.*

Proof: In order to show that S is closer than Q to R we need to show that there is some N such that for $n \geq N$, $d_n(S,R) < d_n(Q,R)$.

Let L be the smallest n such that $P(\omega_n) < Q(\omega_n)$, for some ω_n. Let M be such that for $n \geq M$, $d_n(P,R) \leq d_n(Q,R)$. Take N to be the maximum of L and M. Now, for $n \geq N$,

$$d_n(S,R) = \sum_{\omega_n} [\lambda P(\omega_n) + (1-\lambda)Q(\omega_n)] \log \frac{\lambda P(\omega_n) + (1-\lambda)Q(\omega_n)}{\lambda R(\omega_n) + (1-\lambda)R(\omega_n)}$$

$$< \sum_{\omega_n} \lambda P(\omega_n) \log \frac{\lambda P(\omega_n)}{\lambda R(\omega_n)} + (1-\lambda)Q(\omega_n) \log \frac{(1-\lambda)Q(\omega_n)}{(1-\lambda)R(\omega_n)}$$

$$= \lambda d_n(P,R) + (1-\lambda)d_n(Q,R)$$

$$\leq \lambda d_n(Q,R) + (1-\lambda)d_n(Q,R) = d_n(Q,R).$$

The first inequality is a consequence of the information-theoretic log-sum inequality: $\sum_{i=1}^{k} x_i \log x_i/y_i \geq (\sum_{i=1}^{k} x_i) \log(\sum_{i=1}^{k} x_i)/(\sum_{i=1}^{k} y_i)$ with equality iff x_i/y_i is constant, where $x_1, \ldots, x_k$, and $y_1, \ldots, y_k$ are non-negative real numbers. Here x_i/y_i is not constant because $n \geq L$. The second inequality follows since $n \geq M$. ∎

Corollary 5.4 *If P is closer than Q to R, then any proper convex combination of P and Q is closer than Q to R.*

Note that in the above, P may not be closer to R than convex combination S of P and Q: if P and Q lie either side of R then a proper convex combination may be closer to R than either P or Q. For example, suppose P is defined by $P(\omega_n^+) = 3/4$ and $P(\omega_n^-) = 1/4$, for $\omega_n^+ = A_1 \wedge A_2 \wedge \cdots A_n, \omega_n^- = \neg A_1 \wedge \neg A_2 \wedge \cdots \wedge \neg A_n$, and Q is defined by $Q(\omega_n^+) = 0$ and $Q(\omega_n^-) = 1$. Then, P is closer than Q to the equivocator, but $S = P/2 + Q/2$ sets $S(\omega_n^+) = 3/8$ and $S(\omega_n^-) = 5/8$ and is closer to the equivocator than both P and Q.

On the other hand, if P and Q lie sufficiently on the 'same side' of R, and if as above, P is closer to R than is Q, then P will be closer to R than proper convex combination S. In the extreme case, if P lies on the line segment joining Q and R, then P is a proper convex combination of S and R, so by the log-sum inequality (as in the proof of Theorem 5.3), $d_n(P, R) < d_n(S, R)$ for sufficiently large n.

5.4 Closure

Note that we defined $\downarrow\mathbb{E} \stackrel{\mathrm{df}}{=} \{P \in \mathbb{E} : P \text{ is minimal with respect to } \prec\}$. The question therefore arises as to whether there are in fact any cases in which $\mathbb{E}$ has no minimal elements, that is, elements $P \in \mathbb{E}$ such that for all $Q \in \mathbb{E}, Q \not\prec P$.

Suppose $\mathcal{L}$ is a language with a single unary predicate. Define P_i by

$$P_i(A_j | \omega_{j-1}) = \begin{cases} \frac{1}{2} & : \quad j \leq i \\ 1 & : \quad \text{otherwise} \end{cases}$$

for all j and all $\omega_{j-1} \in \Omega_{j-1}$. Then, we have an infinite descending chain: for all i, $P_{i+1} \prec P_i$. So if $\mathbb{E} = \{P_1, P_2, \ldots\}$, then $\mathbb{E}$ has no minimal elements.

One suggestion for ensuring minimal elements—a suggestion that is *not* taken up in this book for reasons outlined below—is to insist that $\mathbb{E}$ be closed. Perhaps, if $\mathbb{E}$ were augmented by its limit points, then it would have minimal elements after all.

The question of whether this suggestion would work for the above example depends on the notion of limit point that we invoke:

Definition 5.5. (Strong limit point) *Probability function P is a strong limit point of $\mathbb{E}$ if for all $\epsilon > 0$, there is some $Q \neq P$ in $\mathbb{E}$ such that $d_n(Q, P) < \epsilon$, for all n.*

Definition 5.6. (Weak limit point) *Probability function P is a weak limit point of $\mathbb{E}$ if for all $\epsilon > 0$ and for all n, there is some $Q \neq P$ in $\mathbb{E}$ such that $d_n(Q, P) < \epsilon$.*[1]

Now, if we consider the putative characterization of $\mathbb{E}$ as $\{P_1, P_2, \ldots\}$, we see that this set has no strong limit points, but has the equivocator $P_=$ as a weak limit point. Perhaps, if $P_1, P_2, \ldots$ are appropriate candidates for a belief function, then $P_=$ is also an appropriate candidate. In which case, the suggestion would

---

[1] Strictly speaking, Q must be zero wherever P is zero, for $d_n(Q, P)$ to be well defined. We might use Euclidean distance for d_n where this condition does not hold. In fact, since in the context of limit points we are considering small distances, it does not much matter which notion of distance we use in these definitions.

be that we should take the closure of $\mathbb{E}$ under weak limit points as a normative constraint. Then, in fact $\mathbb{E} = \{P_=, P_1, P_2, \ldots\}$ and $\downarrow\!\mathbb{E} = \{P_=\}$.

But there are several reasons why we do *not* invoke closure as a normative constraint.

First, while the closure suggestion ensures a minimal element in the above example, it does not always ensure minimal elements. Again let $\mathcal{L}$ be a language with a single unary predicate. Define Q_i by

$$Q_i(A_j|\omega_{j-1}) = \begin{cases} \frac{1}{2} & : \quad i < j \leq 2i \\ 1 & : \quad \text{otherwise} \end{cases},$$

for all j, ω_{j-1}. This defines an infinite descending chain: for all i, $Q_{i+1} \prec Q_i$. The only weak limit point is Q defined by $Q(A_j|\omega_{j-1}) = 1$, for all j, ω_{j-1}. Hence, in principle there might be some $\mathbb{E} = \{Q, Q_1, Q_2, \ldots\}$. But Q is a maximal element of $\mathbb{E}$: $Q_i \prec Q$, for all i. So $\mathbb{E}$ has no minimal elements.

To take another example, suppose $\mathcal{L}$ has a single unary predicate U and that the evidence $\mathcal{E}$ is just that the A_j form a partition (i.e. the Ut_j are mutually exclusive and exhaustive). Define R_i satisfying $\mathcal{E}$ by

$$R_i(A_j) = \begin{cases} \frac{1}{i} & : \quad j \leq i \\ 0 & : \quad \text{otherwise} \end{cases},$$

for all j. This also defines an infinite descending chain—for all i, $R_{i+1} \prec R_i$—but one with no weak limit points in the set $\mathbb{P}^{\mathcal{L}}$ of probability functions on $\mathcal{L}$. Hence, $\mathbb{E}$ has no minimal elements.

The second reason not to invoke closure is that it is inappropriate given the betting interpretation of belief. Again take $\mathcal{L}$ to be a predicate language with a single unary predicate U such that Ut_i holds iff the ith toss of a coin yields a head uppermost. Suppose the agent knows just that the coin is biased in favour of heads, $\mathcal{E} = \{P^*(Ut_i) > 1/2 : i = 1, 2, \ldots\}$. Taking the characteristic feature of chance to be its connection with frequency, this implies that the limiting relative frequency of U takes some value $x > 1/2$. If a closure principle were adopted, then the agent would be entitled to set $P(Ut_i) = 1/2$, for each i. But placing bets at this value on Ut_i for each i ensures the agent sure loss, as outlined in §3.3. Since x is the limiting relative frequency of U, for any $\epsilon > 0$, there is a natural number N such that for $n \geq N$, $|m/n - x| < \epsilon$, where m is the number of heads uppermost in n tosses. Let $\epsilon = x - 1/2$. The agent's loss on betting on $Ut_1, Ut_2, \ldots, Ut_n$ with betting quotient $\frac{1}{2}$ is $\sum_{i=1}^{n}(1/2 - I_{Ut_i})S_i$, where S_i is the stake on the bet on Ut_i. Now the stake-maker can set each $S_i = -1$ so that her loss is $m - n/2 > n(x - \epsilon) - n/2 = n/2 - n/2 = 0$, for $n \geq N$. Hence, the agent is sure to lose money in the long run.

Note that the convexity condition imposed in §3.3 does not open the agent up to sure loss in this way. If the evidence is that the chance is x or y and the agent bets at rate q where $x < q < y$, then the stake-maker cannot choose stakes guaranteeing sure loss because, having the same evidence as the agent, the stake-maker does not know whether the chance is greater or less than the agent's betting quotient and so does not know whether to set stakes to be positive or negative. Hence, betting considerations tell against closure but not convexity.

The third problem with closure is that there are cases in which, although each probability function in $\mathbb{E}$ satisfies propositional evidence in $\mathcal{E}$, probability functions in the closure of $\mathbb{E}$ do not satisfy this evidence. (Again, this contrasts with convexity: every function in the convex hull of a set of probability functions satisfying θ itself satisfies θ.) To see this, again consider a language $\mathcal{L}$ with a single unary predicate U. Suppose $\mathcal{E} = \{\exists x U(x)\}$. Define $P \notin \mathbb{E}$ by $P(\neg U t_1 \wedge \cdots \wedge \neg U t_n) = 1$, for all n. Define $S_i \in \mathbb{E}$ by

$$S_i(U t_j | \omega_{j-1}) = \begin{cases} 1 & : \quad j = i \\ 0 & : \quad \text{otherwise} \end{cases},$$

for all j, ω_{j-1}. Now, given $\epsilon > 0$ and n there is $S_{n+1} \in \mathbb{E}$ such that $d_n(S_{n+1}, P) = 0 < \epsilon$, so P is a weak limit point of $\mathbb{E}$. But P does not satisfy $\exists x U(x)$.

In sum, then, there are several good reasons not to insist on the closure of $\mathbb{E}$ as a normative constraint on rational belief.

5.5 Characterizing Equivocation

In this section, we consider and reject two possible alternative characterizations of the Equivocation norm.

5.5.1 *Limiting divergence*

Define
$$d(P, Q) \stackrel{\text{df}}{=} \lim_{n \to \infty} d_n(P, Q).$$

As long as P is zero whenever Q is zero, $d(P, Q) \in [0, \infty]$ since for $m < n$,

$$d_n(P, Q) = d_m(P, Q) + \sum_{\omega_n \in \Omega_n} P(\omega_n) \log \frac{P(\omega'_n | \omega_m)}{Q(\omega'_n | \omega_m)} \geq d_m(P, Q) \geq 0,$$

where ω_m is the m-state determined by ω_n and ω'_n is the 'remainder', that is, ω_n is $\omega_m \wedge \omega'_n$.

One might suggest that the agent's degrees of belief should be representable by $P_\mathcal{E} \in \{P \in \mathbb{E} : d(P, P_=) \text{ is minimized}\}$. But as it stands this does not adequately explicate the concept of closeness to the equivocator, because in the case of a predicate language there are probability functions P, Q such that although one is intuitively closer to the equivocator than the other, $d(P, P_=) = d(Q, P_=)$. Suppose, for example, that $\mathcal{E}$ imposes the constraints $P(A_i | \pm A_1 \wedge \cdots \wedge \pm A_{i-1}) = 1$, for all $i \geq 2$ ($\mathbb{E}$ is then non-empty, closed, and convex). Thus, only $P(A_1)$ is unconstrained. Now, $d(P, P_=) = \infty$, for all $P \in \mathbb{E}$. Yet, intuitively there is a unique function in $\mathbb{E}$ that is closest to the equivocator, namely the function that sets $P(A_1) = 1/2$ and $P(A_i | \pm A_1 \wedge \cdots \wedge \pm A_{i-1}) = 1$, for all $i \geq 2$. Indeed, this function is minimal with respect to $\prec$. This motivates taking $P_\mathcal{E} \in \{P \in \mathbb{E} : P \text{ is}$

minimal with respect to $\prec$}, as we did above, rather than $P_\mathcal{E} \in \{P \in \mathbb{E} : d(P, P_=)$ is minimized}.

Example 5.7 Consider a predicate language with a single unary predicate U. Suppose $\mathbb{E} = \{P : P(\forall x U x) = c\}$, for some fixed $c \in [0,1]$. We have that $d(P, P_=) = \infty$, for all $P \in \mathbb{E}$. Define P by

$$P(Ut_1) = \frac{c+1}{2}$$

$$P(Ut_{i+1}|Ut_1 \wedge \cdots \wedge Ut_i) = \frac{(2^{i+1} - 1)c + 1}{(2^{i+1} - 2)c + 2}$$

$$P(Ut_{i+1}|\pm Ut_1 \wedge \cdots \wedge \pm Ut_i) = \frac{1}{2}$$

otherwise. Then P is the member of $\mathbb{E}$ that is closest to the equivocator.

The *closer* relation agrees with comparative limiting divergence in the sense of the following proposition:

Proposition 5.8 $d(P, R) < d(Q, R)$ *implies that P is closer than Q to R.*

Proof: Choose N such that $d_N(Q, R) > d(P, R)$. Then, for $n \geq N$, $d_n(P, R) \leq d(P, R) < d_n(Q, R)$ so P is closer to R than Q. ■

We saw above that the closer relation can distinguish probability functions that cannot be distinguished by limiting divergence—there the limiting divergence was infinite. But the closer relation is also more fine-grained than comparative limiting divergence when limiting divergence is finite. That is, there are P, Q, R such that $d(P, R) = d(Q, R) < \infty$, where P is closer than Q to R: consider, for example, a language with a single unary predicate, $R = P_=$ and P, Q defined by

$$P(A_i|\omega_{i-1}) = \begin{cases} \frac{1}{2} & : i = 1 \\ \frac{1}{2} + \frac{1}{2^{i-1}} & : \text{otherwise} \end{cases}$$

$$Q(A_i|\omega_{i-1}) = \frac{1}{2} + \frac{1}{2^i}.$$

Intuitively, P is more equivocal than Q with respect to A_1 and is equally equivocal otherwise. The closer relation deems P more equivocal than Q, yet limiting divergence fails to distinguish P and W. Hence, the closer relation gives the better ordering, whether or not limiting divergence is infinite.

5.5.2 Pointwise limit of divergence minimizers

There is yet another way in which one might try to characterize the Equivocation norm, which proceeds as follows. Let $\mathbb{E}^n \stackrel{\text{df}}{=} \{P|_{\mathcal{L}_n} : P \in \mathbb{E}\}$, the set of probability functions on $\mathcal{L}_n$ that are restrictions of probability functions (on $\mathcal{L}$) in $\mathbb{E}$. Let P^n be a probability function in $\mathbb{E}^n$ that has minimum n-divergence from the equivocator. (Equivalently, P^n is a probability function in $\mathbb{E}$ that has maximum

n-entropy H_n.) Recall that $\mathbb{E}$ is convex; if $\mathbb{E}$ is also closed, then P^n is uniquely determined: there is a unique minimizer of divergence on a finite domain, from a non-empty, closed convex set of probability functions. Define

$$P^{\infty}(\omega_k) = \lim_{n \to \infty} P^n(\omega_k),$$

for all $\omega_k \in \Omega_k, k = 1, 2, \ldots$. (These values uniquely determine $P(\theta)$ for any sentence $\theta \in S\mathcal{L}$ via PP3–4.) If this leads to a well-defined probability function, one might suggest that one deem P^{∞} to be an appropriate choice for a rational belief function $P_{\mathcal{E}}$.

In fact, where P^{∞} is well defined it *will* often be deemed appropriate by the approach taken here:

Theorem 5.9 *Suppose that P^{∞} exists and is unique. We then have the following chain of implications: $\mathbb{E}$ is closed $\Rightarrow P^{\infty} \in \mathbb{E} \Rightarrow \downarrow\!\mathbb{E} = \{P^{\infty}\}$.*

Proof: [$\mathbb{E}$ is closed $\Rightarrow P^{\infty} \in \mathbb{E}$] First, note that if P^{∞} exists then it defines a probability function:

PP1: $P^{\infty}(\omega_k) = \lim_{n \to \infty} P^n(\omega_k)$ and $P^n(\omega_k) \geq 0$ so $P^{\infty}(\omega_k) \geq 0$.
PP2: If τ is a tautology, then $P^n(\tau) = 1$, for all n, so $P^{\infty}(\tau) = \lim_{n \to \infty} 1 = 1$.
PP3 and PP4 hold trivially in virtue of the way in which P^{∞} is defined.

Next, note that if $\mathbb{E}$ is closed then $P^{\infty} \in \mathbb{E}$ since it is a weak limit point of members of $\mathbb{E}$. One can see this as follows. For each P^n defined as above on $\mathcal{L}_n$, let P_n be some function on $\mathcal{L}$ that extends it and is in $\mathbb{E}$. By definition of P^{∞}, given ω_k and $\epsilon > 0$, $|P^n(\omega_k) - P^{\infty}(\omega_k)| < \epsilon$ for sufficiently large n—say for $n \geq N_{\omega_k}$. Letting $N = \max_{\omega_k \in \Omega_k} N_{\omega_k}$ we see that given $\epsilon > 0$ and $n \geq N$, $|P^n(\omega_k) - P^{\infty}(\omega_k)| < \epsilon$, for all $\omega_k \in \Omega_k$. Equivalently, given $\epsilon > 0$, $d_k(P_n, P^{\infty}) < \epsilon$ for sufficiently large n.[2] But this is just to say that P^{∞} is a weak limit point of the P_n.

[$P^{\infty} \in \mathbb{E} \Rightarrow \downarrow\!\mathbb{E} = \{P^{\infty}\}$] Next, we see that if $P^{\infty} \in \mathbb{E}$, then $P^{\infty} \in \downarrow\!\mathbb{E}$. Suppose for contradiction that $P^{\infty} \notin \downarrow\!\mathbb{E}$. In that case there must be a $Q \in \mathbb{E}$ such that $Q \prec P^{\infty}$. So for sufficiently large n,

$$d_n(P^n, P_=) \leq d_n(Q, P_=) < d_n(P^{\infty}, P_=)$$

so,

$$d_n(P^{\infty}, P_=) - d_n(P^n, P_=) \geq d_n(P^{\infty}, P_=) - d_n(Q, P_=) > 0.$$

Now, $d_n(P^{\infty}, P_=) - d_n(P^n, P_=) \longrightarrow 0$ as $n \longrightarrow \infty$, so $d_n(P^{\infty}, P_=) - d_n(Q, P_=) \longrightarrow 0$ as $n \longrightarrow \infty$. Hence, there is some R such that $d_n(R, P_=) = d_n(P^{\infty}, P_=)$ for all n and $|Q(\omega_n) - R(\omega_n)| \longrightarrow 0$ for each $\omega_n \in \Omega_n$, as $n \longrightarrow \infty$. But the $P(\omega_{n+1})$ for $\omega_{n+1} \in \Omega_{n+1}$ determine the $P(\omega_n)$ for $\omega_n \in \Omega_n$, so it must be the case that $Q = R$. Hence, $d_n(Q, P_=) = d_n(P^{\infty}, P_=)$ for all n, contradicting $Q \prec P^{\infty}$, as required.

---

[2] As noted before, if for some ω_k, $P_n(\omega_k) \neq 0$ but $P^{\infty}(\omega_k) = 0$, then d_k is ill-defined but we can just appeal to Euclidean distance rather than cross-entropy divergence here.

It remains to show that $\downarrow\mathbb{E} = \{P^\infty\}$. Suppose for contradiction that $Q \in \downarrow\mathbb{E}$, for some $Q \neq P^\infty$.

If $d_n(P^\infty, P_=) \leq d_n(Q, P_=)$ for sufficiently large n, then some convex combination S of P^∞ and Q is closer to the equivocator than Q (Theorem 5.3). Since $\mathbb{E}$ is convex, $S \in \mathbb{E}$ and so Q cannot, after all, be in $\downarrow\mathbb{E}$, which provides the required contradiction.

Otherwise, $d_n(Q, P_=) < d_n(P^\infty, P_=)$, for infinitely many n. But by reasoning exactly analogous to that used above to show that $P^\infty \in \downarrow\mathbb{E}$, we have that $d_n(Q, P_=) = d_n(P^\infty, P_=)$ for those n, a contradiction. ∎

On the other hand, if $\mathbb{E}$ is not closed, then P^∞ may not be in $\mathbb{E}$. Consider the sequence P_n as defined at the beginning of §5.4. In this case, $P_n\!\upharpoonright_{\mathcal{L}_n} = P_=^{\mathcal{L}_n}$, the equivocator on $\mathcal{L}_n$, so $P^\infty = P_=$ is well defined. But in that example $P_=$ is not in $\mathbb{E}$.

Now, if $P^\infty \notin \mathbb{E}$ and the arguments of §5.4 against closure are accepted, then P^∞ should not be a candidate for a rational belief function. Hence, Theorem 5.9 can be read as saying that, in cases in which P^∞ is a viable belief function it is deemed so by the framework of §5.1.

Note that Paris and Vencovská (2003) and Barnett and Paris (2008) put forward a rather different notion of limiting belief function. There the procedure is to take pointwise limits of functions that maximize n-entropy from all those satisfying evidence *where the evidence is re-expressed using $\mathcal{L}_n$*. Here the procedure is to take pointwise limits of functions that maximize n-entropy from all those satisfying evidence *expressed using $\mathcal{L}$*. The former case faces the finite model problem: while there may be probability functions that satisfy the evidence on an infinite domain, there may be no probability function that satisfies that evidence when reinterpreted as saying something about a finite domain. This problem arises when considering total orderings, for instance: if the evidence says that $\forall x \exists y Rxy$, where R is a strict total order, then only an infinite language will yield probability functions that satisfy that evidence; hence, one cannot satisfy such a proposition by taking limits of probability functions on finite languages that satisfy the proposition (but one can by taking limits of probability functions that are restrictions of functions on an infinite language that satisfy the proposition). In the approach outlined here, the finite model problem does not arise.

5.6 Order invariance

Next, we turn to the question of whether the *closer* relation is well defined.

There is some flexibility in the ordering of the $A_1, A_2, \ldots$. Although atomic propositions expressible in $\mathcal{L}_m$ must occur before those that are expressible in $\mathcal{L}_n$ but not in $\mathcal{L}_m$, where $m < n$, there will typically be several orderings that satisfy this requirement. However, the *closer* relation is well defined:

Proposition 5.10 *The* closer *relation is independent of the precise ordering $A_1, A_2, \ldots$ of the atomic propositions.*

Proof: Suppose for contradiction that there are two orderings such that P is closer to R than Q under ordering 1 but Q is closer to R than P under ordering 2. Let N_1 and N_2 be the N for orderings 1 and 2, respectively.

Let n be the maximum of N_1 and N_2. Now, $\mathcal{L}_n$ expresses the A_{N_1} of ordering 1 and the A_{N_2} of ordering 2 and all the predecessors of these propositions under both orderings.

Ordering $\mathcal{L}_n$ according to order 1, we have that $d_n(P,R) < d_n(Q,R)$ since $n \geq N_1$. Similarly, if we order $\mathcal{L}_n$ according to order 2, we have that $d_n(Q,R) < d_n(P,R)$ since $n \geq N_2$. But this is a contradiction because cross-entropy divergence is independent of ordering on a finite language. ∎

Note that we have assumed a fixed ordering of the constant symbols in the language. Although *closer* is well defined, the question arises as to how this relation behaves on languages that differ only with respect to the ordering of the t_i.

There is much agreement across orderings, as can be seen from the following consideration. Call two orderings *commensurable* if, given any constant symbol, there is a finite set of constants that contains that constant and is closed with respect to taking ancestors under each ordering; an analogue of the proof of Proposition 5.10 shows that *closer* is independent of ordering where the orderings under consideration are commensurable.

On the other hand, in certain cases the *closer* relation does depend on the ordering of the t_i. Suppose, for instance, that $\mathcal{L}$ and $\mathcal{L}'$ have a single predicate U which is unary, but that in $\mathcal{L}$ the constants are ordered $t_1, t_3, t_2, t_5, t_4, t_7, t_6, t_9, \ldots$ and in $\mathcal{L}'$ they are ordered $t_2, t_4, t_1, t_6, t_3, t_8, t_5, t_{10}, \ldots$. Suppose that P and Q both render the Ut_i all probabilistically independent; let P be defined by $P(Ut_i) = 1$ if i is odd and $1/2$ otherwise, and let Q be defined by $Q(Ut_i) = 1$ if i is even and $1/2$ otherwise. Now, on $\mathcal{L}$ we have that $Q \prec P$, but on $\mathcal{L}'$ we have that $P \prec Q$.

That there is a dependence of probability on this kind of reordering should not be entirely surprising. After all, if we reorder the outcomes of an infinite sequence of coin tosses, we can generate a sequence where the frequency of heads is any value we wish (as long as each of heads and tails occurs infinitely often in the original sequence). This kind of phenomenon is unproblematic because one would have to reorder *by outcome* rather than, say, purely by the times at which the tosses took place. The above example is unproblematic for the same reason.

In our context it is important to point out that this dependence of the *closer* relation on the ordering of the constants does not imply a dependence of the recommendations of objective Bayesian epistemology on the ordering of the constants. The above example yields dependence on ordering because the probability functions are defined in terms of features of the indices of the constants which are not expressible within the language. If the evidence $\mathcal{E}$ imposes finitely many constraints on the probabilities of propositions in $\mathcal{L}$ (as is the case in the probabilistic logic considered in Chapter 7), then $\downarrow\mathbb{E}$ will not be sensitive to the order of the constants. (This is because finitely many constraints can only mention finitely many constant symbols, and hence can only distinguish finitely many atomic propositions. So sensitivity to order can only occur for sufficiently small n. However, the *closer* relation depends on sufficiently large n.)

While this is all well and good for the context of probabilistic logic, the question still remains as to what to do should $\downarrow\mathbb{E}$ be sensitive to order of the constants in other situations. As argued in §9.2, that different agents with different languages adopt different degrees of belief is no problem in itself. What is more problematic is that in natural language there may be no natural order of the constant symbols (i.e. names). If the agent's natural language can be explicated by any one of a number of predicate languages $\mathcal{L}^i$ and rational degree of belief is sensitive to this choice of predicate language, then some protocol is required to handle this sensitivity. The natural protocol is to take $\downarrow\mathbb{E}$ to be $\bigcup_i \downarrow^i\mathbb{E}$, where $\downarrow^i\mathbb{E}$ is the set of probability functions satisfying evidence and closest to the equivocator with respect to language $\mathcal{L}^i$. If there is nothing in the agent's context that determines a most appropriate $\downarrow^i\mathbb{E}$, then there is clearly no rational requirement that the agent's beliefs be representable by a function from one $\downarrow^i\mathbb{E}$ rather than another, and the agent is free to choose among the full range of the $\downarrow^i\mathbb{E}$.

5.7 Equidistance

We have discussed two kinds of indeterminacy: $\downarrow\mathbb{E}$ may not be a singleton due to vagaries of the language (if there is no fixed ordering of the constants), or if there are infinite descending chains (in which case $\downarrow\mathbb{E}$ may be empty—§5.4). But $\downarrow\mathbb{E}$ may contain multiple probability functions simply because it may be the case that there are two or more functions in $\mathbb{E}$ that are closer to the equivocator than other functions in $\mathbb{E}$, but that cannot themselves be ordered by the closeness relation. Such functions are said to be equidistant from the equivocator:

Definition 5.11. (Equidistant) *P and Q are* equidistant *from R if neither is closer than the other to R.*

Proposition 5.12 *For fixed R, the binary relation* equidistant *is reflexive and symmetric but not transitive in general and so not an equivalence relation.*

Proof: Reflexivity and symmetry follow directly from the definitions. To construct a counterexample to transitivity, choose P, Q, and S such that $d_n(P, R)$ and $d_n(Q, R)$ oscillate in magnitude, as do $d_n(Q, R)$ and $d_n(S, R)$, but where $d_n(P, R)$ and $d_n(S, R)$ do not. ∎

Definition 5.13. (Stably Equidistant) *P and Q are* stably equidistant *from R iff there is some N such that for all $n \geq N$, $d_n(P, R) = d_n(Q, R)$.*

Proposition 5.14 *For fixed R,* stably equidistant *is an equivalence relation. If distinct P and Q are stably equidistant from R, then any proper convex combination S of P and Q is closer than either P or Q to R.*

Proof: That *stably equidistant* is an equivalence relation follows directly from the definitions. That S is closer to R than either P or Q follows from Theorem 5.3. ∎

Definition 5.15. (Unstably Equidistant) *P and Q are unstably equidistant from R if they are equidistant from R but not stably equidistant from R.*

Theorem 5.16 *For fixed R, unstably equidistant is irreflexive, symmetric but not transitive in general. P and Q being unstably equidistant from R does not imply that any proper convex combination S of P and Q is closer than either P or Q to R. But a proper convex combination S of P and Q can be no further from R than P or Q. Nor can S and P (respectively S and Q) be stably equidistant from R.*

Proof: Irreflexivity: P and P are *stably* equidistant from R. Symmetry: P and Q are unstably equidistant from R if neither of $d_n(P, R)$ and $d_n(Q, R)$ dominate the other for sufficiently large n, nor are they equal for sufficiently large n; this is clearly symmetric. Failure of transitivity follows from Proposition 5.12.

We show that a convex combination of unstably equidistant probability functions need not be closer to R by providing a counterexample. Take $\mathcal{L}$ to have a single unary predicate, take R to be $P_=$, the equivocator, and define two families P_ζ and Q_ζ of probability functions parameterized by $\zeta \in \mathbb{N}$ as follows. Take the atomic propositions A_i to be probabilistically independent with respect to both P_ζ and Q_ζ. Given n, let $l_\zeta(n) \in \mathbb{N}$ be such that $\zeta^{l_\zeta(n)} \leq n < \zeta^{l_\zeta(n)+1}$. Let

$$P_\zeta(A_n) = \begin{cases} \frac{1}{2} & : \; l_\zeta(n) \text{ is odd} \\ 1 & : \; l_\zeta(n) \text{ is even} \end{cases},$$

$$Q_\zeta(A_n) = \begin{cases} 1 & : \; l_\zeta(n) \text{ is odd} \\ \frac{1}{2} & : \; l_\zeta(n) \text{ is even} \end{cases}.$$

We show first that for sufficiently large ζ, P_ζ and Q_ζ are unstably equidistant. We need to show that $d_i(P, P_=) > d_i(Q, P_=)$, for i in some infinite $I \subseteq \mathbb{N}$, and also that $d_j(P, P_=) < d_j(Q, P_=)$, for $j \in J \subseteq \mathbb{N}$ where J is infinite. Let $I = \{\zeta^{2k+1} - 1 : k \in \mathbb{N}\}$ and $J = \{\zeta^{2k} - 1 : k \in \mathbb{N}\}$.

Recall that, for any probability function T and for $m < n$, if we let

$$d'_n(T, P_=) = \sum_{\omega_n \in \Omega_n} T(\omega_n) \log\left(2^{n-m} T(\omega'_n | \omega_m)\right),$$

then

$$d_n(T, P_=) = d_m(T, P_=) + d'_n(T, P_=) \geq d_m(T, P_=) \geq 0,$$

where ω_m is the m-state determined by ω_n and ω'_n is the remainder, that is, ω_n is $\omega_m \wedge \omega'_n$. Also, $d_m(T, P_=) \leq m \log 2$.

Let $n \in I$ so that $n = \zeta^{2k+1} - 1$ for some k. Let $m = \zeta^{2k} - 1$, so $n - m = (\zeta - 1)\zeta^{2k}$. Consider the atomic states ω_m that P_ζ awards positive probability. P_ζ gives all these r states the same probability, $1/r$. Moreover, for each such state

ω_m there is only one state with positive probability that extends it, namely that in which ω'_n is $A_{m+1} \wedge \cdots \wedge A_n$, that is, ω_n is $\omega_m \wedge A_{m+1} \wedge \cdots \wedge A_n$. Hence,

$$d'_n(P_\zeta, P_=) = \sum_{\omega_m} P_\zeta(\omega_m) \log 2^{n-m} = \frac{r}{r}(n-m)\log 2 = (\zeta - 1)\zeta^{2k} \log 2.$$

So $d_n(P_\zeta, P_=) \geq (\zeta - 1)\zeta^{2k} \log 2$ since $d_m(P_\zeta, P_=) \geq 0$. But $Q_\zeta(\omega'_n|\omega_m) = 2^{-(n-m)}$, so $d'_n(Q_\zeta, P_=) = 0$. Moreover, $d_m(Q_\zeta, P_=) \leq m \log 2 = (\zeta^{2k} - 1)\log 2$. Consequently $d_n(Q_\zeta, P_=) < \zeta^{2k} \log 2$. Hence, if $\zeta \geq 2$, $d_n(P_\zeta, P_=) > d_n(Q_\zeta, P_=)$, for all $n \in I$. Similarly, $d_n(P_\zeta, P_=) < d_n(Q_\zeta, P_=)$, for $n \in J$. Thus, P_ζ and Q_ζ are indeed unstably equidistant for large enough ζ.

Letting $S_\zeta = \frac{1}{2}P_\zeta + \frac{1}{2}Q_\zeta$, we see that S_ζ and Q_ζ are unstably equidistant for sufficiently large ζ (similarly S_ζ and P_ζ). Again take $n \in I$ so that $n = \zeta^{2k+1} - 1$, for some k, and let $m = \zeta^{2k} - 1$, so $n - m = (\zeta - 1)\zeta^{2k}$. Define $\Omega_n^P = \{\omega_n \in \Omega_n : P_\zeta(\omega_n) > 0\}$ and $\Omega_n^Q = \{\omega_n \in \Omega_n : Q_\zeta(\omega_n) > 0\}$. It will be convenient to take Ω_n^P and Ω_n^Q to be disjoint, for all n (this will be used in Equation 5.1): we can achieve this by creating a dummy variable A_0 and stipulating that $P_\zeta(A_0) = 1$ and $Q_\zeta(A_0) = 0$, and that A_0 is independent of all other A_i with respect to both P_ζ and Q_ζ. Note that P_ζ assigns equal probability to each member of Ω_n^P, and similarly for Q_ζ (Equation 5.3). Also, if $\omega_n \in \Omega_n^P$, then $S_\zeta(\omega_n) = P_\zeta(\omega_n)/2 + Q_\zeta(\omega_n)/2 = P_\zeta(\omega_n)/2$; similarly, if $\omega_n \in \Omega_n^Q$, then $S_\zeta(\omega_n) = Q_\zeta(\omega_n)/2$ (Equation 5.2). Now,

$$\begin{aligned}d'_n(S_\zeta, P_=) &= \sum_{\omega_n \in \Omega_n^P} S_\zeta(\omega_n) \log\left(2^{n-m} S_\zeta(\omega'_n|\omega_m)\right) \\ &\quad + \sum_{\omega_n \in \Omega_n^Q} S_\zeta(\omega_n) \log\left(2^{n-m} S_\zeta(\omega'_n|\omega_m)\right) \qquad (5.1) \\ &= \sum_{\omega_n \in \Omega_n^P} \frac{P_\zeta(\omega_n)}{2} \log\left(2^{n-m} \frac{P_\zeta(\omega_n)/2}{P_\zeta(\omega_m)/2}\right) \\ &\quad + \sum_{\omega_n \in \Omega_n^Q} \frac{Q_\zeta(\omega_n)}{2} \log\left(2^{n-m} \frac{Q_\zeta(\omega_n)/2}{Q_\zeta(\omega_m)/2}\right) \qquad (5.2) \\ &= \sum_{\omega_m \in \Omega_m^P} \frac{P_\zeta(\omega_m)}{2} \log\left(2^{n-m} \frac{P_\zeta(\omega_m)}{P_\zeta(\omega_m)}\right) \\ &\quad + \sum_{\omega_m \in \Omega_m^Q} \frac{2^{-(n-m)}Q_\zeta(\omega_m)}{2} \log\left(2^{n-m} \frac{2^{-(n-m)}Q_\zeta(\omega_m)}{Q_\zeta(\omega_m)}\right) \\ &= \frac{1}{2} \log 2^{n-m} + \frac{2^{-(n-m)}}{2} \log\left(2^{n-m} 2^{-(n-m)}\right) \qquad (5.3) \\ &= \frac{n-m}{2} \log 2 \\ &= \frac{(\zeta - 1)\zeta^{2k}}{2} \log 2.\end{aligned}$$

Hence, $d_n(S_\zeta, P_=) \geq \frac{1}{2}(\zeta - 1)\zeta^{2k} \log 2$. But we saw above that $d_n(Q_\zeta, P_=) < \zeta^{2k} \log 2$. So for $n \in I$ and $\zeta \geq 3$, $d_n(S_\zeta, P_=) > d_n(Q_\zeta, P_=)$. Similarly, when $n \in J$, $d_n(S_\zeta, P_=) < d_n(Q_\zeta, P_=)$. Therefore, S_ζ, Q_ζ are unstably equidistant.

In sum, one can show that P and Q being unstably equidistant from R does *not* imply that any proper convex combination S of P and Q is closer than either P or Q to R, by taking $R = P_=$ and P, Q, S to be $P_\zeta, Q_\zeta, S_\zeta$ defined as above for any fixed $\zeta \geq 3$.

It remains to show that convex combination $S = \lambda P + (1-\lambda)Q$ of unstably equidistant P and Q is no further from R than P or Q, and neither are S, P (respectively S, Q) stably equidistant from R. First note that since P and Q are equidistant there are infinite sets I and J of natural numbers such that $d_n(P, R) \leq d_n(Q, R)$ if $n \in I$, and $d_n(P, R) \geq d_n(Q, R)$ if $n \in J$. Since P and Q are unstably equidistant, they are distinct. Then, as shown in the proof of Theorem 5.3, for sufficiently large n

$$d_n(S, R) < \lambda d_n(P, R) + (1-\lambda)d_n(Q, R).$$

Now, if such $n \in I$, $d_n(S, R) < d_n(Q, R)$, so S is no further from R than Q, nor are S and Q stably equidistant from R. Similarly, for large enough $n \in J$, $d_n(S, R) < d_n(P, R)$, so S is no further from R than P, nor are S and P stably equidistant from R. ∎

On the other hand, there exist unstably equidistant P and Q such that some convex combination S *is* closer to R than P or Q: take P and Q either side of R on a line going through R. In fact, by Theorem 5.3, there exist unstably equidistant P and Q such that *every* convex combination S is closer to R than Q: take distinct P and Q such that $d_n(P, P_=) \leq d_n(Q, P_=)$, for all n, with $d_n(P, P_=) = d_n(Q, P_=)$, for infinitely many n and $d_n(P, P_=) < d_n(Q, P_=)$, for infinitely many n.

Proposition 5.17 *Distinct $P, Q \in \downarrow\mathbb{E}$ are unstably equidistant.*

Proof: Since $\downarrow\mathbb{E}$ contains the elements of $\mathbb{E}$ that are minimal with respect to $\prec$, it cannot be the case that $P \prec Q$ or $Q \prec P$. Hence, P and Q are equidistant from the equivocator. If P, Q were stably equidistant from $P_=$ then a convex combination of P and Q would be closer to the equivocator than either P or Q (Proposition 5.14); since $\mathbb{E}$ is convex this would contradict the minimality of P and Q. So P and Q must be unstably equidistant. ∎

Note that $\mathbb{E}$ being convex does not on its own guarantee that $\downarrow\mathbb{E}$ will be convex. Suppose $\downarrow\mathbb{E}$ contains P_ζ, Q_ζ, T where P_ζ and Q_ζ are defined as in the proof of Theorem 5.16 with $\zeta \geq 3$, and T is defined in such a way that

$$d_n(T, P_=) = \min\{d_n(P_\zeta, P_=), d_n(Q_\zeta, P_=)\},$$

for all $n = \zeta^k - 1$ and

$$d_n(T, P_=) < \min\{d_n(P_\zeta, P_=), d_n(Q_\zeta, P_=)\}$$

otherwise. Then, P_ζ, Q_ζ, T are pairwise unstably equidistant from the equivocator, as are P_ζ, Q_ζ, S for $S = P_\zeta/2 + Q_\zeta/2$. But $T \prec S$ since by definition $d_n(T, P_=) < \min\{d_n(P_\zeta, P_=), d_n(Q_\zeta, P_=)\} \leq d_n(S, P_=)$ for all n. Hence, S could not be in $\downarrow\mathbb{E}$ and so $\downarrow\mathbb{E}$ is not convex.

In the case of a propositional language, we saw that $\downarrow\mathbb{E}$ may not be a singleton if $\mathbb{E}$ is not closed. In the case of a predicate language there are now three ways in which $\downarrow\mathbb{E}$ can fail to be a singleton: linguistic indeterminacy, infinite descending chains, and unstable equidistance. While for a propositional language, any closed convex set of probability functions contains a single function that is closest to the equivocator, with a predicate language this need not be the case.[3] Accordingly, in this book we make no presumption that $\downarrow\mathbb{E}$ is a singleton. Non-uniqueness of $P_\mathcal{E} \in \downarrow\mathbb{E}$ poses no conceptual difficulty, but in some applications of Bayesian epistemology—including the application to probabilistic logic considered in Chapter 7—care needs to be taken to keep track of all the potential belief functions. The question of the relationship between non-uniqueness of $P_\mathcal{E}$ and the objectivity of $P_\mathcal{E}$ will be taken up in §9.3.

---

[3] Having said this, preliminary results indicate that $\downarrow\mathbb{E}$ is a singleton in the restricted environment of probabilistic logic on unary languages (Jeff Paris and Soroush Rad, personal communication).

6

Objective Bayesian Nets

Objective Bayesianism has been criticized on the grounds that representing and reasoning with maximum entropy probability functions appears to be computationally infeasible for all but small problems (see, e.g. Pearl 1988, p. 463). In this chapter, we develop computational machinery that permits efficient representation and inference for objective Bayesian probability. This is the machinery of *objective Bayesian nets* (obnets) and *objective credal nets*.

Bayesian and credal nets are introduced in §6.1. They are adapted to handle objective Bayesian probability in §6.2. Then, by way of example, the formalism is applied to the problem of cancer prognosis in §6.3.

6.1 Probabilistic networks

Consider a language $\mathcal{L}_n = \{A_1, \ldots, A_{r_n}\}$; this might be a propositional language on propositional variables $A_1, \ldots, A_n$ (in which case $r_n = n$) or a finite predicate language with constants $t_1, \cdots, t_n$ and atomic propositions $A_1, \ldots, A_{r_n}$ (in which case $r_n \geq n$). In §§2.3 and 5.1, we saw that a probability function on such a language is determined by its values on the atomic n-states $\omega_n \in \Omega_n$. There are 2^{r_n} of these n-states. Since the probabilities of the n-states must sum to 1, any one of these probability values is determined by the others. It is plain to see that as n increases the number $2^{r_n} - 1$ of probability values to be specified increases exponentially. It seems plausible, then, that n would not have to become very large for representing probability functions and reasoning with probability functions to become intractable.

The apparatus of probabilistic networks is used to render probabilistic representation and reasoning more tractable. The key feature of probabilistic networks is that they offer a *dimension reduction*. While the space $\mathbb{P}$ of probability functions on $\mathcal{L}_n$ has dimension $2^{r_n} - 1$, in the sense that, in general, that many coordinates $P(\omega_n)$ are required to specify an arbitrary probability function, a probabilistic network exploits probabilistic independence relations satisfied by a particular probability function to reduce the number of coordinates required to specify the function. In many applications of probability, the probability functions of interest typically satisfy so many independence relationships (and are hence specifiable by so few coordinates) that they can be handled quite tractably.

6.1.1 *Bayesian nets*

A *Bayesian net* is one kind of probabilistic network, used to represent a single probability function. Bayesian nets were developed to represent probability functions defined over sets of variables (Pearl 1988; Neapolitan 1990), but they

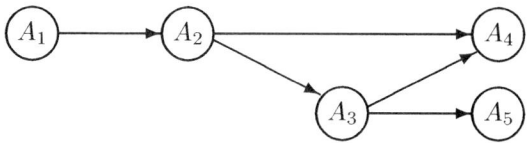

FIG. 6.1. A directed acyclic graph (dag).

TABLE 6.1. The probability distribution of A_4 conditional on A_2 and A_3.

$P(\neg A_4\|\neg A_2 \wedge \neg A_3) = 0.7$	$P(A_4\|\neg A_2 \wedge \neg A_3) = 0.3$
$P(\neg A_4\|\neg A_2 \wedge A_3) = 0.9$	$P(A_4\|\neg A_2 \wedge A_3) = 0.1$
$P(\neg A_4\|A_2 \wedge \neg A_3) = 0.2$	$P(A_4\|A_2 \wedge \neg A_3) = 0.8$
$P(\neg A_4\|A_2 \wedge A_3) = 0.4$	$P(A_4\|A_2 \wedge A_3) = 0.6$

can equally be applied to represent probability functions over logical languages, which are our concern here. Suppose P is a probability function defined on $\mathcal{L}_n$. A Bayesian net representation of P consists of a *directed acyclic graph* (or *dag* for short) whose nodes are $A_1, \ldots, A_{r_n}$, together with the conditional probability distribution $P(A_i|Par_i)$, induced by P, of each A_i conditional on its parents Par_i in the graph. For example, Fig. 6.1 depicts a dag; a conditional probability distribution of A_4 conditional on A_2 and A_3 is specified in Table 6.1 (note that by the additivity of probability, only four of these probabilities, one on each row of the table, need to be specified). The graph must be constructed in such a way that the following condition holds:

Markov Condition: Each A_i is probabilistically independent of its non-descendants in the graph, conditional on its parents, written $A_i \perp\!\!\!\perp ND_i \mid Par_i$.

For example, the graph of Fig. 6.1 implies that

$$A_3 \perp\!\!\!\perp A_1 \mid A_2.$$
$$A_4 \perp\!\!\!\perp A_1, A_5 \mid A_2, A_3.$$
$$A_5 \perp\!\!\!\perp A_1, A_2, A_4 \mid A_3.$$

A Bayesian net determines a probability function P on $\mathcal{L}_n$ via the identity

$$P(\omega_n) = \prod_{i=1}^{r_n} P(A_i^{\omega_n}|Par_i^{\omega_n}),$$

where $A_i^{\omega_n}$ and $Par_i^{\omega_n}$ are the states of A_i and its parents that are consistent with ω_n. In the above example,

$$P(A_1 \wedge \neg A_2 \wedge \neg A_3 \wedge A_4 \wedge A_5) =$$

$$P(A_1)P(\neg A_2|A_1)P(\neg A_3|\neg A_2)P(A_4|\neg A_2 \wedge \neg A_3)P(A_5|\neg A_3).$$

Every probability function is representable by a Bayesian net. If the function satisfies no probabilistic independencies, one can simply take a Bayesian net on

TABLE 6.2. A constrained set of probability distributions of A_4 conditional on A_2 and A_3.

$P(\neg A_4|\neg A_2 \wedge \neg A_3) \in [0.7, 0.8]$	$P(A_4|\neg A_2 \wedge \neg A_3) \in [0.2, 0.3]$
$P(\neg A_4|\neg A_2 \wedge A_3) = 0.9$	$P(A_4|\neg A_2 \wedge A_3) = 0.1$
$P(\neg A_4|A_2 \wedge \neg A_3) \in [0.2, 1]$	$P(A_4|A_2 \wedge \neg A_3) \in [0, 0.8]$
$P(\neg A_4|A_2 \wedge A_3) \in [0.35, 0.4]$	$P(A_4|A_2 \wedge A_3) \in [0.6, 0.65]$

a complete dag (one in which there is an arrow between every pair of nodes) since such a graph implies no independencies via the Markov condition.

The advantage of a Bayesian net representation of a probability function is that typically fewer numbers need to be specified in order to fully specify a probability function. Instead of having to specify $P(\omega_n)$ for all but one $\omega_n \in \Omega_n$ (a total of 31 probabilities in our example), we only need specify the values in the probability tables—a total of 11 probabilities. This dimension reduction becomes crucial when n is large.

6.1.2   Credal nets

A *credal net* is another kind of probabilistic network, used to represent a set of probability functions. As with a Bayesian net, a credal net invokes a dag. But instead of specifying, for each propositional variable / atomic proposition A_i, a single distribution of A_i conditional on its parents, it specifies constraints that such a probability distribution must satisfy. It does this by associating a closed interval—rather than a single real number—with each state of A_i and its parents (though single numbers, as trivial closed intervals, are still permitted). Consider a credal net on the dag of Fig. 6.1; an example of a probability table is given in Table 6.2 (again by additivity of probability we require only one entry on each row to determine the whole table).

The Markov Condition is also applied to a credal net. The Markov Condition can be interpreted in various ways when referring to a set of probability functions (Haenni *et al.* 2010, §8.1.1), but here we appeal to the following interpretation: for *each* probability function represented by the network, each A_i is probabilistically independent of its non-descendants in the dag, conditional on its parents, written $A_i \perp\!\!\!\perp_P ND_i \mid Par_i$, for all $P \in \mathbb{C}$, the set of probability functions determined by the credal net. Under this construal, the credal net can be thought of as the set of Bayesian nets that have the graph of the credal net and whose conditional probability distributions satisfy the constraints in the credal net.

A credal net then determines a set $\mathbb{C}$ of probability functions on $\mathcal{L}_n$, via $\mathbb{C} = \{P : P$ satisfies the constraints in the tables and $P(\omega_n) = \prod_{i=1}^{r_n} P(A_i^{\omega_n}|Par_i^{\omega_n})\}$, where $A_i^{\omega_n}$ and $Par_i^{\omega_n}$ are the states of A_i and its parents that are consistent with ω_n.

But not every set of probability functions is representable by a credal net as defined above. For a start, only closed sets of functions are representable.

$$A_1 \longrightarrow A_2$$

Fig. 6.2. A graph for the set $\mathbb{S}$.

Table 6.3. A probability table for A_1.

$P(A_1) = x \in [0.3, 1]$

Moreover, thus far the constraints in the probability tables are treated as operating independently, but to represent certain sets of probability functions we need the value P takes within one constraining interval to restrict the value it can take within another. For example, consider $\mathcal{L}_2$ (with $r_n = n$) and the set of probability functions $\mathbb{C} = \{P : P(A_1 \wedge A_2) = 0.3, P(A_2|\neg A_1) = 0\}$. Let us try to build a credal net representing this set on the graph (Fig. 6.2), which implies no independencies via the Markov condition. To lie in this set, a probability function P must satisfy $P(A_1) \in [0.3, 1]$, which yields a probability table for A_1. Moving on to the probability table for A_2, P must also satisfy $P(A_2|\neg A_1) = 0$, but the value $P(A_2|A_1)$ takes depends on the value of $P(A_1)$: $P(A_2|A_1) = 0.3/x$, where $x = P(A_1)$. So we would need to represent the constraints via Table 6.3 and 6.4.

If we link the probability tables by appealing to parameters such as x above, we have a *parameterized credal net*. Appealing to parameters increases the expressive power of credal nets: parameters enable credal nets to represent a wider range of sets of probability functions.

While specifying a credal net requires specifying more numbers than specifying a Bayesian net—both end points of each closed interval—credal nets afford the same dimension reduction as Bayesian nets. Depending on the sparsity of the dag, the number of constraints in the probability tables can be far fewer than the $2^{r_n} - 1$ that would be required to constrain all but one $P(\omega_n)$ individually.

6.1.3 *Inference*

Having specified a probabilistic net, it is natural to want to infer $P(\psi)$, or a range of possible values for $P(\psi)$, from the net, where ψ is some sentence of $\mathcal{L}_n$. There are a wide range of algorithms for performing inference in probabilistic networks, most of which are specialized to certain kinds of inference (e.g. to inferring certain conditional probabilities from Bayesian nets). One very general approach is that of Haenni (2007) and Haenni *et al.* (2010, §8.2) which has the advantage that it treats Bayesian nets, credal nets, and parameterized credal

Table 6.4. A probability table for A_2 conditional on A_1.

$P(A_2|\neg A_1) = 0$
$P(A_2|A_1) = 0.3/x$

nets in a homogeneous way, and which can also handle arbitrary propositional sentences ψ. This approach, which we now briefly sketch, combines a compilation methodology with numerical approximation methods.

The approach works by splitting inference into two phases. The first, *compilation* phase transforms the probabilistic net into a new representation (a 'deterministic, decomposable negation normal form' network) which permits very efficient inference. The second, *query-answering* phase performs inference in the new representation. The first phase is computationally expensive but needs to be carried out only once, while the second phase is computationally cheap and can be carried out each time a new ψ is presented.

We should note that inference with credal nets can be computationally much more complex than inference with Bayesian nets: the credal net may represent a set of probability functions with a large number of extremal points, each of which needs to be kept track of to perform inference. The compilation methodology outlined above helps mitigate this computational burden, but it is also helpful to use numerical approximation methods, rather than exact inference methods, when working with credal nets. Accordingly, hill-climbing numerical methods for compiled credal nets are well suited. These methods work perfectly well in the extreme case in which the credal net is in fact a Bayesian net; since we may not know in advance whether this case will arise, this is a distinct advantage.

Here is a high-level summary of the compilation algorithm:

Algorithm 6.1

  Input: A credal net on $\mathcal{L}_n$ and a sentence ψ of $\mathcal{L}_n$.

  - Compilation phase:
    * Transform the credal net into a deterministic Decomposable Negation Normal Form (d-DNNF) net.
  - Inference phase:
    * Transform ψ into $\psi_1 \vee \cdots \vee \psi_l$, where the ψ_i are mutually exclusive conjunctions of literals—for example, via Abraham's algorithm (Abraham 1979).
    * Use hill-climbing in the d-DNNF net to calculate approximate bounds on each $P(\psi_i)$.
    * Calculate bounds on $P(\psi)$ via the identity $P(\psi) = \sum_{i=1}^{l} P(\psi_i)$.

  Output: $Y = \{P(\psi) : P \text{ is subsumed by the input credal net}\}$.

6.2 Representing objective Bayesian probability

6.2.1 *Objective Bayesian nets*

An *obnet* is a Bayesian net representation of a function $P_\mathcal{E}$ on $\mathcal{L}_n$ that, according to objective Bayesian epistemology, yields appropriate degrees of belief for an agent with language $\mathcal{L}_n$ and evidence $\mathcal{E}$.

An obnet can be constructed by (i) determining independencies that must be satisfied by $P_\mathcal{E}$, (ii) representing these independencies by a dag that satisfies the Markov Condition, and then (iii) determining the conditional distributions $P_\mathcal{E}(A_i | Par_i)$.

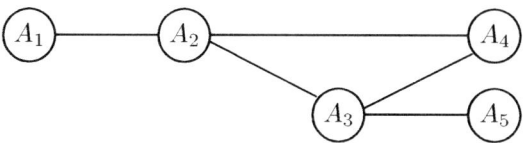

Fig. 6.3. Constraint graph.

Task (i) is straightforward as we now see. We denote by χ the set of chance constraints—the constraints imposed by quantitative information in $\mathcal{E}$ that bears on chances. We delay discussion of structural constraints until the end of this section, supposing for the moment that $\mathbb{E} = \langle \mathbb{P}^* \rangle = \{P : P \text{ satisfies } \chi\}$.

Definition 6.2. (Constraint Graph) *The* constraint graph *for $\mathcal{E}$ on $\mathcal{L}_n$ is constructed by taking the $A_1, \ldots, A_{r_n}$ as nodes and linking two nodes with an edge if they occur in the same constraint in χ. (In the case in which $\mathcal{L}_n$ is a predicate language and constraints involve quantifiers, first substitute each occurrence of $\forall x \theta(x)$ by $\bigwedge_t \theta(t)$; similarly, substitute each occurrence of $\exists x \theta(x)$ by $\bigvee_t \theta(t)$.)*

Theorem 6.3 *For all $P_\mathcal{E} \in \downarrow\mathbb{E} = \downarrow\langle \mathbb{P}^* \rangle$, separation in the constraint graph implies conditional independence in $P_\mathcal{E}$: for $X, Y, Z \subseteq \{A_1, \ldots, A_{r_n}\}$, if Z separates X from Y in the constraint graph, then $X \perp\!\!\!\perp Y \mid Z$ for $P_\mathcal{E}$.*

Proof: See Williamson (2005a, theorem 5.1). ∎

For example, suppose the constraint set takes the form $\chi = \{P(A_1 \wedge \neg A_2) \in [0.8, 0.9], P((\neg A_4 \vee A_3) \to A_2) = 0.2, P(A_5) + P(A_3) \in [0.3, 0.6], P(A_4) = 0.7\}$. Then, the constraint graph of Fig. 6.3 represents conditional independencies that must be satisfied by any maximally equivocal function satisfying these constraints. For instance, A_2 separates A_1 from A_3, A_4, A_5, so A_1 must be probabilistically independent of $\{A_3, A_4, A_5\}$ conditional on A_2, for each probability function satisfying the constraints that has maximum entropy.

Step (ii), representing these independencies by a dag, can be performed by the following algorithm[1]:

Algorithm 6.4

Input: An undirected graph $\mathcal{G}$.

- Triangulate $\mathcal{G}$ to give $\mathcal{G}^T$.
- Reorder the variables according to maximum cardinality search.
- Let $D_1, \ldots, D_l$ be the cliques of $\mathcal{G}^T$, ordered according to highest labelled vertex.
- Let $E_j = D_j \cap (\bigcup_{i=1}^{j-1} D_i)$ and $F_j = D_j \setminus E_j$, for $j = 1, \ldots, l$.
- Construct a dag $\mathcal{H}$ by taking propositional variables as nodes, and
  * Add an arrow from each vertex in E_j to each vertex in F_j, for $j = 1, \ldots, l$.

---

[1] See, for example, Neapolitan (1990) for the graph theoretic terminology.

FIG. 6.4. V: virtuous; H: (hu)man; M: mortal; s: Socrates.

* Add further arrows, from lower numbered variables to higher numbered variables, to ensure that there is an arrow between each pair of vertices in $D_j, j = 1, \ldots, l$.

Output: A dag $\mathcal{H}$.

When input a constraint graph this algorithm produces a dag that satisfies the Markov Condition with respect to any maximally equivocal function compatible with this evidence (Williamson 2005a, theorem 5.3). A graph produced by this algorithm typically looks much like the constraint graph that is input: for example, when input Fig. 6.3, Fig. 6.1 is one possible output of the algorithm.

Step (iii) in the construction of an obnet is the determination of the conditional distributions $P(A_i|Par_i)$. Here any of a variety of techniques (for instance, numerical methods or Lagrange multiplier methods) can be used to determine values of these parameters that maximize entropy subject to constraints imposed by evidence. If $\mathbb{E}$ is closed—for example, if each constraint is of the form $\sum_{i=1}^{k} x_i P(\theta_i) \geq y$ or $\sum_{i=1}^{k} x_i P(\theta_i|\varphi_i) \geq y$—then since $\mathcal{L}_n$ is finite there is a unique entropy maximizer. Hence these conditional distributions are uniquely determined. We do not explore the determination of these conditional distributions in any detail in this book; the importance of a Bayesian net lies in dimension reduction and hence in the graph rather than in the numerical parameters.

Consider another simple example. Suppose that χ constrains the agent to believe *all men are mortal* $(\forall x, H(x) \to M(x))$ to degree at least 3/4, *all those who are virtuous are men* $(\forall x, V(x) \to H(x))$ to degree at least 3/5, and *Socrates is virtuous* (Vs) to degree 4/5. The graph of a resulting obnet is depicted in Fig. 6.4. The corresponding probabilities are

$$P(Vs) = \frac{4}{5};$$

$$P(Hs|Vs) = \frac{3}{4}, P(Hs|\neg Vs) = \frac{1}{2};$$

$$P(Ms|Hs) = \frac{5}{6}, P(Ms|\neg Hs) = \frac{1}{2}.$$

It turns out then that the agent should believe that Socrates is mortal to degree $P(Ms) = 11/15$.

6.2.2 Objective credal nets

In cases where $P_{\mathcal{E}}$ is not uniquely determined, we may be interested in representing not only an agent's chosen $P_{\mathcal{E}}$ but indeed all candidates for $P_{\mathcal{E}}$, that is, we may want to represent $\Downarrow\mathbb{E}$ (or indeed $\Downarrow E$). In such cases, a single obnet does not suffice to represent $\Downarrow\mathbb{E}$. Note, however, that thanks to Theorem 6.3 all members of $\downarrow\mathbb{E}$ may be represented by obnets on the same graph. Hence, if $\downarrow\mathbb{E}$ is closed and convex, it may be representable by a credal net. An *objective credal net* (or *ocnet* for short) is a credal net that represents the set $\downarrow\mathbb{E}$ that is of fundamental interest to objective Bayesianism. The graph of this ocnet can be determined by Definition 6.2 and Algorithm 6.4.

6.2.3 Infinitary nets

Bayesian nets and obnets are thus far defined over a finite language $\mathcal{L}_n$. But in this book we also consider the case in which the agent's language $\mathcal{L}$ is an infinite predicate language. Now, as we saw in Chapter 5, an infinite language $\mathcal{L}$ can be handled naturally by considering the sequence $\mathcal{L}_1, \mathcal{L}_2, \ldots$ of finite languages generated by the ordering of the constant symbols: each $\mathcal{L}_n$ involves only constants $t_1, \ldots, t_n$. A probability function P over $\mathcal{L}$ is determined by its values over the $\mathcal{L}_n(n = 1, 2, \ldots)$, so it is determined by the sequence of Bayesian nets representing P over the $\mathcal{L}_n(n = 1, 2, \ldots)$. These Bayesian nets can be constructed in such a way that each net contains its predecessor in the sequence as a strict subnet; hence, we can define the corresponding *infinitary Bayesian net* to be the Bayesian net on the infinite domain $A_1, A_2, \ldots$ that has each member of the sequence of (finite) Bayesian nets as a subnet. In turn, then, $P_\mathcal{E} \in \downarrow\mathbb{E}$ may be represented by a sequence of obnets, or an infinitary obnet. Similarly, $\downarrow\mathbb{E}$ itself can often be represented by a sequence of ocnets or an *infinitary credal net*. In practice, of course, one will at any stage only be interested in propositions involving the symbols of some fixed $\mathcal{L}_n$, so one can restrict attention to the finite subnet on that particular $\mathcal{L}_n$.

6.2.4 Qualitative evidence

We now extend the obnet approach to handle the structural constraints imposed by qualitative evidence.

Recall from the discussion of §3.3 that qualitative evidence about influence relations—for example, causal, logical, semantic, and hierarchical relationships—constrains degrees of belief in a way that is not mediated by $\mathbb{P}^*$, the set within which the chance function is suspected to lie. There we explicated $\mathbb{E}$, the set of probability functions compatible with evidence, as $\langle\mathbb{P}^*\rangle \cap \mathbb{S}$ where $\mathbb{S}$ is the set of structural constraints imposed by qualitative evidence. We saw that structural constraints take the form $P_\mathcal{E}^\mathcal{L}|_{\mathcal{L}'} = P_{\mathcal{E}'}^{\mathcal{L}'}$, where $\mathcal{L}'$ is a sublanguage of $\mathcal{L}$ and $\mathcal{E}'$ is the evidence concerning propositions expressible in that sublanguage.

Again let χ be the chance constraints imposed by *quantitative* evidence. The constraint graph is constructed, as before, from just these chance constraints. It turns out that admitting structural constraints has no effect on the crucial separation property of Theorem 6.3, which depends just on the chance constraints:

Theorem 6.5 *For all $P_\mathcal{E} \in \downarrow\mathbb{E} = \downarrow(\langle\mathbb{P}^*\rangle \cap \mathbb{S})$, separation in the constraint graph implies conditional independence in $P_\mathcal{E}$: for $X, Y, Z \subseteq \{A_1, \ldots, A_{r_n}\}$, if Z separates X from Y in the constraint graph, then $X \perp\!\!\!\perp Y \mid Z$ for $P_\mathcal{E}$.*

Proof: See Williamson (2005a, theorem 5.6). ∎

This key fact means that the construction of the graph in the objective Bayesian/credal net proceeds exactly as before: build the constraint graph from chance constraints and then use Algorithm 6.4 to convert this into a dag that is guaranteed to satisfy the Markov Condition. The structural constraints are required only when it comes to determining the probability tables of the net.

In fact, structural constraints simplify the determination of the net. If there is a single structural constraint $P_{\mathcal{E}}^{\mathcal{L}} \downarrow_{\mathcal{L}'} = P_{\mathcal{E}'}^{\mathcal{L}'}$, then one can split the problem of determining the net into two sub-problems: first construct a net on $\mathcal{L}'$ from evidence $\mathcal{E}'$, then extend this net to one on $\mathcal{L}$ using evidence $\mathcal{E}$. If there are more structural constraints, the determination of the net becomes an iterative process. In the limiting case in which the agent knows all the influence relationships, the dag representing the influence relationships (e.g. the causal graph) can feature as the dag in the obnet, and the probability tables are determined in sequence, reducing an entropy maximization problem over the whole net to r_n entropy maximization problems each involving one node and its parents; see Williamson (2005a, §5.8) for a detailed discussion of this case.

6.3 Application to cancer prognosis

Having presented the obnet formalism, we now sketch how this formalism can be used in practice. We consider the application of obnets to breast cancer prognosis, described in detail in Nagl *et al.* (2008). The problem here is that a patient has breast cancer and an agent—a medical practitioner—must make an appropriate treatment decision. Some treatments have harsh side effects and it would not be justifiable to inflict these on low-risk patients. Broadly speaking, the higher the probability of recurrence of the cancer, the more aggressive the treatment that should be given. So it is important to determine the degree to which the medical practitioner should believe the patient's cancer will recur.

Here the available evidence is multifarious and exhibits various kinds of structures. Evidence includes the following. There are a variety of clinical data sets describing the clinical symptoms and disease progress of past patients. There are genomic data sets describing the presence or absence of molecular markers in past patients. There are scientific papers that provide evidence of causal relations and mechanisms, and that provide statistical information which quantifies the strength of connection between the variables under study. Causal relationships and mechanisms can also be elicited from experts in the field, such as clinicians and cancer biology researchers. And there are also a whole host of prior medical informatics systems which provide a variety of evidence: for example, evidence of ontological relationships between variables in medical ontologies and evidence of logical relationships in medical argumentation systems.

Traditional machine learning methodology would take one of two standard courses. One option is to choose the best piece of data—for example, a clinical data set—and to build a model (e.g. a Bayesian net) that represents the distribution of that data. The resulting model would then be used as a basis for decision. Clearly this approach ignores much of the available evidence, and will not yield useful results if the chosen data is not plentiful, accurate, and relevant. A second option is to build a model from each piece of evidence and to combine the results—for example, by each model taking a vote on the recommended decision and somehow aggregating these votes. There are several difficulties with this approach. One is that most machine learning methods only take a data set as input; consequently, the qualitative causal evidence and the evidence concerning hierarchical mechanisms is likely to be ignored. A second is that the resulting models may be based on mutually inconsistent assumptions, in which case it is not clear that they should be combined at all. A third difficulty is that the problem of aggregating the judgements of the various models is itself fraught (Chapter 8). In contrast, the approach taken in Nagl et al. (2008) is to construct a *single* model—an obnet—that takes into account the full range of evidence. This study considered four evidential sources, which are now described.

The first source is the SEER study, a clinical data set involving 3 million patients in the United States from 1975 to 2003; of these 4,731 were breast cancer patients. This data set measures the following variables: age, tumour size (mm), grade (1–3), HR Status (oestrogen/progesterone receptors), lymph node tumours, surgery, radiotherapy, survival (months), status (alive/dead). A sample of the data set appears below:

Age	T Size	Grade	HR	LN	Surgery	Radiotherapy	Survival	Status
70–74	22	2	1	1	1	1	37	1
45–49	8	1	1	0	2	1	41	1
...	...	...	...	...	...	...	...	...

If standard machine learning methods for learning a Bayesian net that represents the empirical probability distribution of this data set were invoked, they would generate a net with a graph similar to that of Fig. 6.5. In our case, however, we treat this empirical distribution as a constraint on appropriate degrees of belief. An agent's degree of belief in any proposition that involves only variables measured in this data set should match the empirical probability of that proposition as determined by the data set: $P_{\mathcal{E}}(\theta) = P^*(\theta)$ for all θ involving just variables in the data set.

The second source of evidence consists of genomic data from a Progenetix data set, with 502 cases. A sample appears below:

1p31	1p32	1p34	2q32	3q26	4q35	5q14	7p11	8q23	20p13	Xp11	Xq13
0	0	0	1	-1	0	0	1	0	0	0	-1
0	0	1	1	0	0	0	-1	-1	0	0	0
...	...	...	...	...	...	...	...	...	...	...	...

The empirical distribution of this data set is represented by a Bayesian net with the graph of Fig. 6.6. Again, from an objective Bayesian point of view, this data imposes the constraint that $P_{\mathcal{E}}(\theta) = P^*(\theta)$, for all θ involving just variables in the data set.

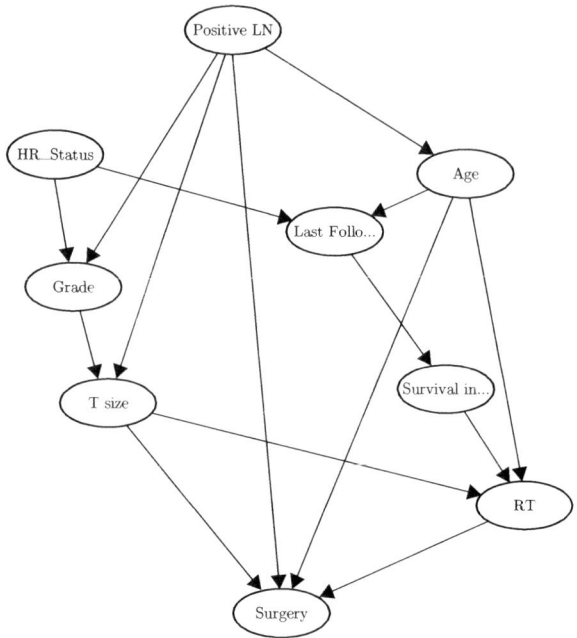

FIG. 6.5. Graph of a Bayesian net representing the empirical distribution of the clinical data.

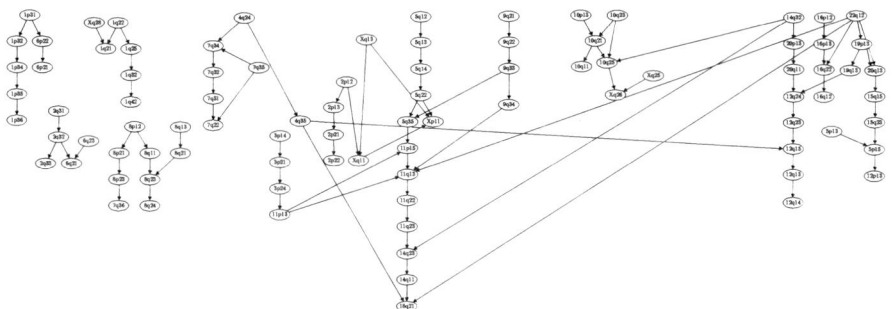

FIG. 6.6. Graph in a Bayesian net representation of a genomic data set.

The third source was a further genomic data set (119 cases with clinical annotation) from the Progenetix database:

Lymph Nodes	1q22	1q25	1q32	1q42	7q36	8p21	8p23	8q13	8q21	8q24
0	1	1	1	1	0	0	0	0	0	0
1	0	0	0	0	0	0	0	0	0	0
...	...	...	...	...	...	...	...	...	...	...

The fourth source was a published study (Fridlyand et al. 2006), which contains causal and quantitative information concerning the probabilistic

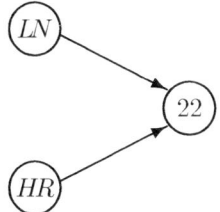

FIG. 6.7. Lymph nodes status, hormone receptor status, and 22q12.

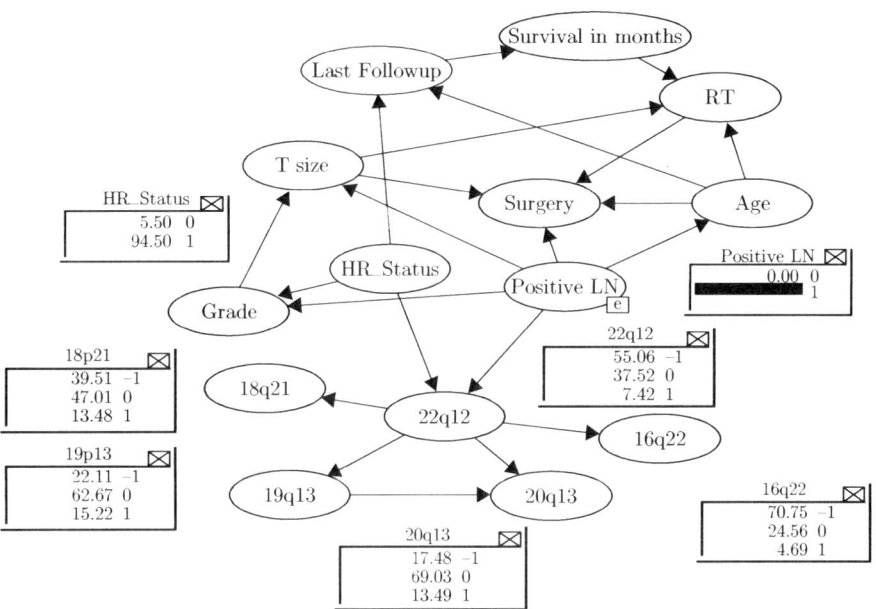

FIG. 6.8. Graph of the objective Bayesian net (obnet).

dependence between the variables HR_status and 22q12—this provided a further bridge between clinical and genomic variables represented in Fig. 6.7.

The resulting obnet has the graph depicted in Fig. 6.8. (Here variables with little relevance to prognosis have been removed.)

This kind of representation is attractive in that it involves both clinical and molecular variables, permitting inferences from one kind of variable to the other. Thus, one can use molecular as well as clinical evidence to determine an appropriate prognosis. Bridging the molecular and clinical levels is one of the goals of systems biology and obnets offer a principled way of achieving this goal. This kind of approach is also attractive in that it encapsulates the three norms of objective

Bayesianism: cancer prognosis requires probabilistic predictions that are based on all available evidence and that only commit to outcomes to the minimum extent forced by evidence; because of their basis in objective Bayesianism, obnets fulfil this need (Russo and Williamson 2007).

7

Probabilistic Logic

In this chapter, we see that objective Bayesianism can be applied as a semantics for probabilistic logic. Section 7.1 presents the rather general framework for probabilistic logic of Haenni *et al.* (2010). Since this is a purely formal framework, it is open to a number of different interpretations. In §7.2, we give a flavour of some of these interpretations, turning to the objective Bayesian interpretation in §7.3. We see that in general credal nets can provide a calculus for probabilistic logic; in particular, objective credal nets provide a calculus for objective Bayesian probabilistic logic (§7.4).

7.1 A formal framework for probabilistic logics

7.1.1 *Entailment*

Broadly speaking, there are three main kinds of probabilistic logic, or *progic* for short. In an *internal* progic, a logical language $\mathcal{L}$ includes function symbols that are interpreted as probability functions. One then performs standard logical inference in order to draw conclusions that say something about these probability functions. An internal progic—where probabilities are internal to the language—is thus useful for reasoning *about* uncertainty (Halpern 2003). In contrast, in an *external* progic the logical language $\mathcal{L}$ does not involve probabilities but instead probabilities are ascribed to the sentences of $\mathcal{L}$—the probabilities are external to the language. One then performs probabilistic inference to draw conclusions that say something about the probabilities of sentences. An external progic is thus used for reasoning *under* uncertainty (Paris 1994). The third case is that of a *mixed* progic, which contains probabilities internal to the language and external to the language: a mixed progic can be used to reason about and under uncertainty.

In order to be fully general, we consider a mixed progic in this chapter. But for clarity of exposition we consider a mixed progic of a fixed structure; the discussion of this chapter can be generalized to other forms of mixed progic, or indeed particularized to internal and external progics.

Let $\mathcal{L}$ be a propositional or predicate language of the kind considered in previous sections. (Recall that we considered only predicate languages with no function symbols, so there are no probabilities internal to $\mathcal{L}$.) We are primarily interested in inferences of the following form:

$$\varphi_1^{X_1}, \ldots, \varphi_k^{X_k} \approx \psi^Y. \tag{7.1}$$

Here $\varphi_1, \ldots, \varphi_k, \psi$ are sentences of $\mathcal{L}$, and $X_1, \ldots, X_k, Y$ are sets of real numbers (which we will construe as probabilities) that attach to these sentences. $\approx$ is an unspecified entailment relation; we are more specific about this relation in subsequent sections of this chapter.

Thus far the probabilities are external to $\mathcal{L}$. Note, however, that the entailment relation $\approx$ relates sentences together with attached probabilities. So the premises and conclusion are expressions in some language richer than $\mathcal{L}$. Accordingly, consider a propositional language $\mathcal{L}^\sharp$ whose propositional variables are expressions of the form φ^X where φ is a sentence of $\mathcal{L}$ and X is a set of probabilities. (Note that as it stands there are uncountably many propositional variables, though one could always circumscribe the language $\mathcal{L}^\sharp$ to reduce its cardinality.) By applying the usual connectives of propositional logic, we can generate sentences μ of $\mathcal{L}^\sharp$ of arbitrary logical complexity. The metalanguage $\mathcal{L}^\sharp$ can be thought of as the language of the entailment relation $\approx$, and this entailment relation can be extended to inferences of the more general form

$$\mu_1, \ldots, \mu_k \approx \nu, \tag{7.2}$$

where $\mu_1, \ldots, \mu_k, \nu$ are arbitrary sentences of $\mathcal{L}^\sharp$. Here the probabilities are internal to the language $\mathcal{L}^\sharp$, though they are external to the language $\mathcal{L}$. Thus, the entailment relation $\approx$ yields a progic that is, in a sense, mixed; it can be used to reason both about and under uncertainty.

Clearly, Equation 7.1 and 7.2 might be interpreted in a variety of ways according to one's understanding of the uncertainties and the entailment relation involved. We deem the resulting logic to be a *probabilistic* logic if the X_i, Y are treated as sets of probabilities and if the logic in question invokes probability functions as models, that is, if the logic deems the entailment relation to hold just when every probability function that satisfies the premises on the left-hand side of the entailment symbol $\approx$ also satisfies conclusion on the right-hand side, where 'satisfies the premises' and 'satisfies the conclusion' are given suitable explications. We use the notation $[\![\mu_1, \ldots, \mu_k]\!]$ to denote the set of probability functions that satisfy $\mu_1, \ldots, \mu_k$ treated as premises, and we use $P \approx \nu$ to assert that probability function P satisfies ν treated as a conclusion. In a probabilistic logic, then, $[\![\cdots]\!]$ and $\approx$ must be interpreted in such a way that $\mu_1, \ldots, \mu_k \approx \nu$ holds iff $P \approx \nu$ for each probability function $P \in [\![\mu_1, \ldots, \mu_k]\!]$. In §7.3, we develop an interpretation that is motivated by objective Bayesian epistemology, but we consider others in §7.2.

7.1.2 Inference

Non-probabilistic logic faces an inferential question of the following form: given premises and conclusion, do the premises entail the conclusion? A *proof* is normally constructed to answer this kind of question, though truth tables and semantic trees can sometimes provide alternative calculi.

But a progic of the sort outlined above more typically faces a question of a rather different form: given premises $\varphi_1^{X_1}, \ldots, \varphi_n^{X_k}$ and a conclusion sentence

ψ of $\mathcal{L}$, what set of probabilities Y should attach to ψ? This question can be represented thus,

$$\varphi_1^{X_1}, \ldots, \varphi_n^{X_k} \approx \psi^? \tag{7.3}$$

More generally, such a question may take the form

$$\mu_1, \ldots, \mu_k \approx \psi^?, \tag{7.4}$$

where $\mu_1, \ldots, \mu_k$ are arbitrary sentences of $\mathcal{L}^\sharp$ and as before ψ is a sentence of $\mathcal{L}$. Since these are primarily questions about probability, answering them will tend to invoke techniques from probability theory rather than proof theory.

In certain cases—for example, if the set $[\![\mu_1, \ldots, \mu_k]\!]$ of probability functions that satisfies the premises is a closed convex set—credal nets (§6.1) can be used as a means to answer questions of the above form. The idea is to build a credal net that represents $[\![\mu_1, \ldots, \mu_k]\!]$ and then to perform inference in this net to determine the smallest set Y such that $P \approx \psi^Y$, for each $P \in [\![\mu_1, \ldots, \mu_k]\!]$. Usually, we will have that $Y = \{P(\psi) : P \in [\![\mu_1, \ldots, \mu_k]\!]\}$, in which case, although the credal net depends on the interpretation of $[\![\cdots]\!]$ (i.e. depends on the chosen semantics for the probabilistic logic), the method for calculating Y will be independent of the chosen semantics. In §6.1, we saw that compilation techniques can be used to fulfil this role, but other methods are of course possible.

The credal net approach is then to probabilistic logic what proof is to non-probabilistic logic. Just as axioms and rules of inference depend on the particular non-probabilistic logic in question, so too does the credal net depend on the particular probabilistic logic. But once the axioms and rules are provided the proof methodology is independent of particular non-probabilistic logic; similarly, once the method for constructing the credal net is provided then the inference methodology is independent of particular probabilistic logic.

7.2 A range of semantics

In this section, we will give a flavour of the array of particular probabilistic logics that fit into the above framework. In §7.3, we turn to the objective Bayesian variant.

A probabilistic logic deems the entailment relation $\varphi_1^{X_1}, \ldots, \varphi_k^{X_k} \approx \psi^Y$ to hold iff every probability function P that satisfies the premises satisfies the conclusion. So the primary task of a semantics for probabilistic logic is to say what 'satisfies the premises' and 'satisfies the conclusion' means.

7.2.1 Standard semantics

According to the standard probabilistic semantics 'satisfies the premises' and 'satisfies the conclusion' are both given the standard interpretation: P satisfies φ^X iff $P(\varphi) \in X$. We write $P \models \varphi^X$ for this satisfaction relation; $P \models \varphi^X$ iff $P \in \mathbb{P}_{\varphi^X} = \{P \in \mathbb{P} : P(\varphi) \in X\}$. So if we define

$$P \approx \psi^Y \Leftrightarrow P \models \psi^Y,$$

$$[\![\varphi_1^{X_1},\ldots,\varphi_k^{X_k}]\!] = \mathbb{P}_{\varphi_1^{X_1},\ldots,\varphi_k^{X_k}},$$

then $\varphi_1^{X_1},\ldots,\varphi_k^{X_k} \approx \psi^Y$ holds iff $P \approx \psi^Y$ for every probability function $P \in [\![\varphi_1^{X_1},\ldots,\varphi_k^{X_k}]\!]$. Similarly, $\mu_1,\ldots,\mu_k \approx \nu$ holds iff the set of probability functions satisfying the premises is a subset of those satisfying the conclusion, $\mathbb{P}_{\mu_1,\ldots,\mu_k} \subseteq \mathbb{P}_\nu$.

Note that the premises may have no satisfiers, $[\![\mu_1,\ldots,\mu_k]\!] = \emptyset$, in which case the entailment relation holds trivially. Note too that even if $X_1,\ldots,X_k$ are all singletons, Y may not be: for example, $A \wedge B^{.9} \approx A^{[.9,1]}$ for this semantics (here for brevity we omit the set brackets $\{\}$ around a single probability). But on a propositional language if the X_i are all closed intervals, then the set $[\![\mu_1,\ldots,\mu_k]\!]$ of functions satisfying the premises is closed and convex and the minimal Y that can attach to the conclusion is also a closed interval, $Y = [P(\psi), Q(\psi)]$ where P and Q are vertices of the set of probability functions $[\![\mu_1,\ldots,\mu_k]\!]$. For example,

$$A \wedge B^{.25}, A \vee \neg B^1 \approx A^{[.25,1]},$$

$$A \wedge B^{.25}, A \vee \neg B^1 \approx B^{.25},$$

$$A \wedge B^{[0,.25]}, A \vee \neg B^1 \approx A^{[0,1]}.$$

Inference for the standard semantics would normally be a very large linear programming problem (Nilsson 1986). However, where the premises incorporate probabilistic independence statements the credal net approach becomes useful. Accordingly, we extend $\mathcal{L}^\sharp$ to include conditional independence statements of the form $\theta \perp\!\!\!\perp \varphi \mid \psi$, where $\theta, \varphi, \psi \in S\mathcal{L}$, in order to be able to express entailment questions such as

$$C^{.8}, A \wedge \neg C^{.1}, A \perp\!\!\!\perp B \mid C \approx A \vee B^?.$$

The graph of the credal net can be constructed by starting with a complete undirected graph, removing edges to capture as many of the given independencies as possible, and then orienting the remaining edges appropriately. This achieves a dimension reduction. The constraints on the conditional probability distributions can then be determined by solving a system of linear equations (Haenni et al. 2010, §9.2).

7.2.2 Probabilistic argumentation

The aim of probabilistic argumentation is to model argumentation under uncertainty (Haenni 2009). The core idea is that an argument should, together with what is already known, force the truth of the conclusion. The *degree of support* for a sentence ψ is a measure of the credibility of those arguments for ψ: it is the probability that ψ is provable.

Given a question
$$\varphi_1^{X_1}, \varphi_2^{X_2}, \ldots, \varphi_k^{X_k} \approx \psi^?.$$
where $\varphi_1, \ldots, \varphi_k, \psi$ are sentences of a propositional language $\mathcal{L}$, we can separate the premises into those that are uncertain and those that are certain:
$$\varphi_1^{X_1}, \varphi_2^{X_2}, \ldots, \varphi_j^{X_j}, \varphi_{j+1}^1, \ldots, \varphi_k^1 \approx \psi^?,$$
where $X_1, \ldots, X_j \neq \{1\}$. The *probabilistic variables* are the propositional variables in $\varphi_1, \ldots, \varphi_j$. Let $\mathcal{L}'$ be the set of probabilistic variables, and consider the set Ω' of atomic states of $\mathcal{L}'$. The *arguments* in favour of ψ are defined to be those states of $\mathcal{L}'$ that, together with the certain premises, logically imply the conclusion sentence: $Args(\psi) = \{\omega' \in \Omega' : \omega' \wedge \varphi_{j+1} \wedge \cdots \wedge \varphi_k \models \psi\}$. Let α_φ be the sentence $\bigvee Args(\varphi)$. The *conflicts*, $Args(\bot)$, are the states of $\mathcal{L}'$ that are inconsistent with the certainties $\varphi_{j+1}, \ldots, \varphi_k$ (and we set $\alpha_\bot \stackrel{\text{df}}{=} \bigvee Args(\bot)$). The *non-conflicts* are the remaining states of $\mathcal{L}'$, represented by the sentence $\neg\alpha_\bot \stackrel{\text{df}}{=} \bigvee(\Omega' \setminus Args(\bot))$. Define the *degree of support* for ψ, given probability function P on $\mathcal{L}'$, by
$$dsp_P(\psi) = P(\alpha_\psi | \neg \alpha_\bot) = \frac{P(\alpha_\psi) - P(\alpha_\bot)}{1 - P(\alpha_\bot)}.$$
The degree of support for ψ is the probability that a non-conflict is an argument in favour of ψ. We can then answer the question by setting
$$Y = \{dsp_P(\psi) : P \models \varphi_1^{X_1}, \varphi_2^{X_2}, \ldots, \varphi_j^{X_j}\}.$$
Note that degree of support, although defined on $\mathcal{L}$, is not a probability function on $\mathcal{L}$: $dsp_P(\psi)$ is sub-additive as a function of ψ.

We can rewrite Y as $Y = \{P(\alpha_\psi) : P \in \mathbb{P}^{\mathcal{L}'}, P \models \neg\alpha_\bot, \alpha_{\varphi_1}^{X_1}, \alpha_{\varphi_2}^{X_2}, \ldots, \alpha_{\varphi_j}^{X_j}\}$. We see then that according to the semantics provided by probabilistic argumentation, $\varphi_1^{X_1}, \varphi_2^{X_2}, \ldots, \varphi_k^{X_k} \approx \psi^Y$ holds iff every probability function P on $\mathcal{L}'$ satisfying the premises also satisfies the conclusion, that is, iff $P \approx \psi^Y$ for each $P \in [\![\varphi_1^{X_1}, \varphi_2^{X_2}, \ldots, \varphi_k^{X_k}]\!]$, where
$$P \approx \psi^Y \Leftrightarrow P \models \alpha_\psi^Y,$$
$$[\![\varphi_1^{X_1}, \varphi_2^{X_2}, \ldots, \varphi_j^{X_j}, \varphi_{j+1}^1, \ldots, \varphi_k^1]\!] = \{P \in \mathbb{P}^{\mathcal{L}'} : P \models \neg\alpha_\bot, \alpha_{\varphi_1}^{X_1}, \alpha_{\varphi_2}^{X_2}, \ldots, \alpha_{\varphi_j}^{X_j}\}.$$

Probabilistic argumentation acts as a kind of bridge between classical logic and probabilistic logic under the standard semantics. If $j = 0$ probabilistic argumentation behaves like classical logic, while if $j = k$ it yields the standard semantics (Haenni 2009, §3.6). But in other cases it yields a nonmonotonic logic (see Definition 7.3), and, from a formal point of view, probabilistic argumentation corresponds very closely to the Dempster–Shafer theory of belief functions (Haenni 2009, §3.7).

Since models are probability functions on $\mathcal{L}'$, inference in probabilistic logic under the probabilistic argumentation semantics can be performed by building a credal network whose nodes are propositional variables of $\mathcal{L}'$. (As usual the set of models must be representable by a credal net; this holds, for instance, if the $X_1, \ldots, X_j$ are closed intervals.) As before, we can extend the language $\mathcal{L}'$ to admit premises that assert probabilistic independencies. However, some versions of the probabilistic argumentation formalism assume that the propositional variables of $\mathcal{L}'$ are all probabilistically independent (Haenni et al. 2000, §2.4), in which case the dag in the credal net is discrete (i.e. contains no arrows). In general, the dag can take into account both independencies that arise from the premises and those that are encapsulated in the semantics as assumptions. The graph construction method is the same as that of the standard semantics.

7.2.3 Evidential probability

Evidential probability considers inferences made on the basis of evidence of statistical frequencies (see, e.g. Kyburg Jr and Teng 2001). The theory of evidential probability considers questions

$$\varphi_1^{X_1}, \varphi_2^{X_2}, \ldots, \varphi_k^{X_k} \mathrel{\vert\approx} \psi^?,$$

where $\varphi_1, \ldots, \varphi_k, \psi$ are sentences of a predicate language $\mathcal{L}$ that includes statistical statements of the form $\%x(\theta(x), \rho(x), [l, u])$, signifying that the frequency of attribute θ in reference class ρ is in the interval $[l, u]$, that is, that between l and u ρs are also θs. The first step in an evidential probability inference is to discard those premises $\varphi_i^{X_i}$ for which $X_i \not\subseteq [1 - \delta, 1]$, for some preordained risk level δ. The remaining $\varphi_{i_1}, \ldots, \varphi_{i_l}$ are called the *evidential certainties*.

Evidential probability has a number of rules for drawing inferences from the statistical statements amongst the evidential certainties. Roughly put, one statistical statement trumps another if it is based on a measurement of more variables (called the *richness* principle), if it measures a narrower reference class (*specificity*), or if its interval is narrower (*strength*). Applying these tests in order leaves a set of untrumped statistical statements relevant to sentence $\theta(t)$ (relevant in the sense that it is known that the reference class applies to t, that is, that $\rho(t)$). The theory of evidential probability then associates a particular interval with $\theta(t)$, namely the smallest closed interval that contains all the intervals of these relevant statistics. This interval is the *evidential probability* of $\theta(t)$. If the interval attaching to $\theta(t)$ lies within $[1 - \epsilon, 1]$ for some preordained ϵ, then $\theta(t)$ is called a *practical certainty* and is inferred.

Suppose for example that $\delta = \epsilon = 0.1$. Then, the machinery of evidential probability would yield the following inference:

$$\%x(A(x), B(x) \wedge C(x), [.95, .96])^{.8}, \%x(A(x), B(x), [.95, .98])^{[.99, 1]}, Bt \wedge Ct^1 \mathrel{\vert\approx} At.$$

The first premise is discarded because it is not an evidential certainty, $.8 < 1 - \delta$. The remaining premises can be used to infer that $P(At) \in [.95, .98]$. Since $.95 \geq 1 - \epsilon$, At itself is then inferred as a practical certainty.

Evidential probability as formulated by Henry Kyburg is not probabilistic in the usual sense of the point-valued probability functions discussed in this book; rather, evidential probabilities are intervals. Since we are concerned in this chapter with probabilistic logics in the usual sense of probability, we appeal to a modification of evidential probability that handles point-valued probabilities (Wheeler and Williamson 2010). This modification, called *second-order evidential probability*, also differs from Kyburg's theory in some other respects. First, it considers all the premises, whether or not they make the grade as evidential certainties. Second, it takes the uncertainties X_i attaching to the premises into account in its calculations. Third, it is concerned, not with inferring practical certainties, but with determining the uncertainty that attaches to a conclusion sentence. One might ask, for example,

$$\%x(A(x), B(x) \wedge C(x), [.95, .96])^{.8},$$

$$\%x(A(x), B(x), [.95, .98])^{[.99, 1]}, Bt \wedge Ct^1 \approx P(At) \in [.95, .98]^?.$$

Taking all the premises and their uncertainty levels into account, what probability should attach to the conclusion sentence?

According to second-order evidential probability, $\varphi_1^{X_1}, \varphi_2^{X_2}, \ldots, \varphi_n^{X_n} \approx \psi^Y$ holds iff $P(\psi) \in Y$ for all probability functions P such that (i) $P(\varphi_1) \in X_1, \ldots, P(\varphi_n) \in X_n$. (ii) P respects the rules of evidential probability for handling statistical statements (i.e. if given Φ the evidential probability of θ is $[l, u]$, then $P(\theta|\Phi) \in [l, u]$). (iii) P is distributed uniformly over this evidential probability interval $[l, u]$, unless premises imply otherwise (i.e. if given Φ the evidential probability of θ is $[l, u]$, then $P(P(\theta(t)) \in [l', u']|\Phi) = \frac{|[l, u] \cap [l', u']|}{|[l, u]|}$). (iv) P renders the premise sentences φ_i independent, unless premises imply otherwise. This yields a genuine probabilistic logic: models are probability functions in the usual sense, P 'satisfies the premises' means (i)–(iv), and P 'satisfies the conclusion' just means $P \models \psi^Y$:

$$P \approx \psi^Y \Leftrightarrow P \models \psi^Y,$$

$$[\![\varphi_1^{X_1}, \varphi_2^{X_2}, \ldots, \varphi_j^{X_n}]\!] = \{P \in \mathbb{P}^\mathcal{L} : P \models \varphi_1^{X_1}, \varphi_2^{X_2}, \ldots, \varphi_j^{X_n}, \text{(ii–iv) hold}\}.$$

Inference in second-order evidential probability can be performed using credal nets. The structure of evidential probability calculations determine the structure of the credal net. To take a very simple example, consider the question

$$\%x(\theta(x), \rho(x), [.2, .4])^{[.9, 1]}, \rho(t)^1 \approx P(\theta(t)) \in [.2, .4]^?.$$

Letting φ_1 be $\%x(\theta(x), \rho(x), [.2, .4])$, φ_2 be $\rho(t)$, and ψ be $P(\theta(t)) \in [.2, .4]$, the dag for the credal net is depicted in Fig. 7.1. The premises are independent in virtue of (iv).

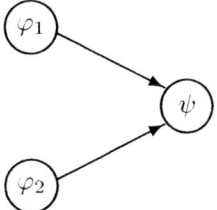

FIG. 7.1. A dag for a second-order evidential probability credal net.

The X_i and the evidential probability inferences determine the conditional probability constraints via (i)–(iii) above:

$$P(\varphi_1) \in [.9, 1], P(\varphi_2) = 1,$$

$$P(\psi|\varphi_1 \wedge \varphi_2) = 1, P(\psi|\neg\varphi_1 \wedge \varphi_2) = .2 = P(\psi|\varphi_1 \wedge \neg\varphi_2) = P(\psi|\neg\varphi_1 \wedge \neg\varphi_2).$$

We then find that $P(\varphi) \in [.92, 1]$, so

$$\%x(\theta(x), \rho(x), [.2, .4])^{[.9,1]}, \rho(t)^1 \approx P(\theta(t)) \in [.2, .4]^{[.92,1]}.$$

7.2.4 Bayesian statistics

Bayesian statistical inference has the following distinctive features. First, probabilities are attached to statistical hypotheses: statistical hypotheses are formulated in the language over which a probability function P is defined, and this function ascribes *higher-order* probabilities since a statistical hypothesis may itself be thought of as one or more probability functions ascribing probabilities to events. Second, Bayes' theorem is widely used to calculate the probability of a hypothesis θ conditional on evidence ε:

$$P(\theta|\varepsilon) = \frac{P(\varepsilon|\theta)P(\theta)}{P(\varepsilon)}.$$

Since θ itself typically determines or heavily constrains the likelihoods $P(\varepsilon|\theta)$, the contentious quantities tend to be $P(\theta)$ and $P(\varepsilon)$, the prior probability of the hypothesis and evidence, respectively.

Given a question

$$\varphi_1^{X_1}, \varphi_2^{X_2}, \ldots, \varphi_k^{X_k} \approx \psi^?,$$

we can separate the premises into those that are uncertain and those that are certain

$$\varphi_1^{X_1}, \varphi_2^{X_2}, \ldots, \varphi_j^{X_j}, \varphi_{j+1}^1, \ldots, \varphi_k^1 \approx \psi^?,$$

where $X_1, \ldots, X_j \neq \{1\}$. We can interpret the certainties as the evidence, setting ε to be $\varphi_{j+1} \wedge \cdots \wedge \varphi_j$. We can also interpret the uncertainties $\varphi_1^{X_1}, \varphi_2^{X_2}, \ldots, \varphi_j^{X_j}$

as information about the prior probabilities. Suppose hypothesis ψ imposes constraints $\chi_1,\ldots,\chi_l$ on the likelihoods $P(\cdot|\psi)$, then we can answer the question by setting

$$Y = \{P(\psi|\varepsilon) : P \models \varphi_1^{X_1},\ldots,\varphi_j^{X_j}, \chi_1,\ldots,\chi_l\}.$$

This gives another probabilistic logic: models are probability functions and $\varphi_1^{X_1}, \varphi_2^{X_2},\ldots,\varphi_k^{X_k} \approx \psi^Y$ holds iff every probability function P that satisfies the premises also satisfies the conclusion, where 'satisfies the premises' means in this case that $P \models \varphi_1^{X_1},\ldots,\varphi_j^{X_j}, \chi_1,\ldots,\chi_l$ and 'satisfies the conclusion' means that $P(\psi|\varphi_{j+1} \wedge \cdots \wedge \varphi_k) \in Y$:

$$P \approx \psi^Y \Leftrightarrow P(\psi|\varphi_{j+1} \wedge \cdots \wedge \varphi_k) \in Y,$$

$$[\![\varphi_1^{X_1},\varphi_2^{X_2},\ldots,\varphi_j^{X_j},\varphi_{j+1}^1,\ldots,\varphi_k^1]\!] = \{P : P \models \varphi_1^{X_1},\ldots,\varphi_j^{X_j}, \chi_1,\ldots,\chi_l\}.$$

Again we can use credal nets for inference. In this case we have a three-step process. First, build a credal net representing $\{P : P(\varphi_1) \in X_1,\ldots,P(\varphi_j) \in X_j, \chi_1,\ldots,\chi_l\}$ (assuming as usual that the X_i are such that this set can be represented by such a net). This net can be built using the same methods that were employed in the case of the standard probabilistic semantics. Second, update the credal net on evidence $\varphi_{j+1},\ldots,\varphi_k$. (A wide variety of algorithms are available for updating—that is, performing Bayesian conditionalization on—a probabilistic network. See Neapolitan 1990 for an introduction.) Third, use the common inference machinery for calculating $P(\psi)$ for each P represented by the updated net.

This approach to probabilistic logic extends Bayesian statistical inference to cases in which there is no fully specified prior, and to cases without fully specified likelihoods (e.g. with interval-valued likelihoods).

7.3 Objective Bayesian semantics

We have seen that entailment relationships in probabilistic logic can be interpreted via the standard probabilistic semantics, probabilistic argumentation, evidential probability or Bayesian statistical inference. We now see that they can also be interpreted by appealing to objective Bayesian epistemology.

Given a question of the form of Equation 7.3—that is, $\varphi_1^{X_1},\ldots,\varphi_k^{X_k} \approx \psi^?$, for sentences $\varphi_1,\ldots,\varphi_k,\psi$ of $\mathcal{L}$—first interpret $\mathcal{L}$ as the language of an agent, and the premises $\varphi_1^{X_1},\ldots,\varphi_k^{X_k}$ as characterizing her evidence. Thus, the premises are construed as facts about empirical probability determined by the agent's evidence $\mathcal{E}$: a premise of the form φ^X is interpreted as $P^*(\varphi) \in X$, where P^* is chance, and the evidence $\mathcal{E}$ is taken to imply that $P^*(\varphi_1) \in X_1,\ldots,P^*(\varphi_k) \in X_k$. As long as the premises are consistent, the Calibration principle then deems the set $\mathbb{E} = \langle \mathbb{P}^* \rangle$ of probability functions that are compatible with this evidence to be the convex hull of the set $\mathbb{P}^* = \{P \in \mathbb{P}_\mathcal{L} : P(\varphi_1) \in X_1,\ldots,P(\varphi_k) \in X_k\}$. (If the premises are inconsistent, then, as discussed in §3.3, a consistency maintenance procedure needs to be invoked to determine $\mathbb{P}^*$. For concreteness, we take maximal consistent subsets here: $\mathbb{P}^* = \{P \in \mathbb{P}_\mathcal{L} : P \models \chi, \chi \in \biguplus\{\varphi_1^{X_1}, \ldots, \varphi_k^{X_k}\}\}$, where $\biguplus\{\varphi_1^{X_1}, \ldots, \varphi_k^{X_k}\}$ is the set of maximal consistent subsets of $\{\varphi_1^{X_1}, \ldots, \varphi_k^{X_k}\}$.)

We can now go on to interpret a question of the form $\varphi_1^{X_1}, \ldots, \varphi_k^{X_k} \approx \psi^?$ as asking how strongly an agent whose evidence is characterized by the premises should believe the conclusion sentence ψ. The objective Bayesian answer is that the agent's degree of belief in ψ should be representable by the probability $P_\mathcal{E}(\psi)$ where the agent's belief function $P_\mathcal{E} \in \Downarrow\!\mathbb{E}$, that is, $P_\mathcal{E}$ is a sufficiently equivocal probability function from all those compatible with the premises. Thus, the optimal set Y of probabilities to attach to the conclusion sentence is $Y = \{P(\psi) : P \in \Downarrow\!\langle\mathbb{P}^*\rangle\}$.

It is straightforward to extend this approach to handle questions of the form of Equation 7.4, $\mu_1, \ldots, \mu_k \approx \psi^?$. As before the premises are interpreted as statements about empirical probability and $\mathbb{P}^* = \{P \in \mathbb{P}_\mathcal{L} : P \models \mu_1, \ldots, \mu_n\}$. As before, the optimal answer to the question is $Y = \{P(\psi) : P \in \Downarrow\!\langle\mathbb{P}^*\rangle\}$.

In turn, we can interpret an entailment claim of the form of Equation 7.1—that is, $\varphi_1^{X_1}, \ldots, \varphi_k^{X_k} \approx \psi^Y$—as holding iff an agent with evidence characterized by the premises $\varphi_1^{X_1}, \ldots, \varphi_k^{X_k}$ should believe the conclusion sentence ψ to some degree within the set Y. Thus, $\varphi_1^{X_1}, \ldots, \varphi_k^{X_k} \approx \psi^Y$ iff $\{P(\psi) : P \in \Downarrow\!\langle\mathbb{P}^*\rangle\} \subseteq Y$. More generally, we can consider an entailment claim of the form of Equation 7.2—that is, $\mu_1, \ldots, \mu_k \approx \nu$ where $\mu_1, \ldots \mu_k, \nu$ are sentences of $\mathcal{L}^\sharp$—to hold iff the conclusion ν must hold of the degrees of belief of an agent with evidence characterized by the premises $\mu_1, \ldots, \mu_k$.

This objective Bayesian interpretation yields a genuine probabilistic entailment relation in the sense that models are probability functions and the entailment relation holds iff all probability functions P that satisfy the premises also satisfy the conclusion. Here P satisfies the premises iff $P \in \Downarrow\!\langle\mathbb{P}^*\rangle$, where $\mathbb{P}^*$ is determined by the premises as outlined above. P satisfies the conclusion iff it satisfies it in the standard sense, $P \models \psi^Y$. In sum,

$$P \approx \psi^Y \Leftrightarrow P \models \psi^Y,$$

$$[\![\mu_1, \ldots, \mu_k]\!] = \Downarrow\!\langle\{P \in \mathbb{P}_\mathcal{L} : P \models \chi, \chi \in \biguplus\{\mu_1, \ldots, \mu_k\}\}\rangle.$$

7.3.1 Sufficiently equivocal probability functions

Hitherto in our discussion of objective Bayesianism, pragmatic, and contextual factors have played the role of determining which probability functions are *sufficiently* equivocal, $\Downarrow\!\mathbb{E}$. However, a question written in the sparse form $\varphi_1^{X_1}, \ldots, \varphi_k^{X_k} \approx \psi^?$ offers little in the way of contextual guidance and there is a danger that objective Bayesianism will fall silent as to the set $Y = \Downarrow\!\mathbb{E}$. Note that as a minimum we have the constraints mentioned at the end of §2.3: $\downarrow\!\mathbb{E} \subseteq \Downarrow\!\mathbb{E} \subseteq \mathbb{E}$ and $\Downarrow\!\mathbb{E} \neq \emptyset$. So in order to say something concrete, it will be useful to adopt the following default identification:

$$\Downarrow\!\mathbb{E} = \begin{cases} \downarrow\!\mathbb{E} & : \downarrow\!\mathbb{E} \neq \emptyset \\ \mathbb{E} & : \text{otherwise} \end{cases}.$$

unless contextual factors decree otherwise. The stipulation that $\Downarrow\mathbb{E} = \mathbb{E}$ if $\downarrow\mathbb{E} = \emptyset$ is adopted for the following reason. If $\downarrow\mathbb{E} = \emptyset$, then that is because there are infinite descending chains, that is, sequences $P_1, P_2, \ldots$ of probability functions in $\mathbb{E}$ such that each P_{i+1} is closer to the equivocator than P_i. Now, if we were to take $\Downarrow\mathbb{E}$ to be a strict subset $\mathbb{F}$ of $\mathbb{E}$ in cases with infinite descending chains, then idempotence could fail, $\Downarrow\Downarrow\mathbb{E} \neq \Downarrow\mathbb{E}$. (If $\mathbb{F}$ is an infinite strict subset of an infinite descending chain, then $\mathbb{F}$ itself would contain an infinite descending chain. On the other hand, if $\mathbb{F}$ were a finite subset of an infinite descending chain, then if $\mathbb{E}$ contained infinitely many infinite descending chains, the union $\bigcup \mathbb{F}_i$ of the finite subsets could itself contain an infinite descending chain. Applying the $\Downarrow$ operator again to an infinite descending chain would yield yet another strict subset, so $\Downarrow\Downarrow\mathbb{E} \neq \Downarrow\mathbb{E}$.) But idempotence is intuitively very plausible when applied to 'sufficiently': a function that is sufficiently close to the equivocator, from all those that are sufficiently close to the equivocator, is clearly itself sufficiently close to the equivocator, so $\Downarrow\Downarrow\mathbb{E} \subseteq \Downarrow\mathbb{E}$; on the other hand, if a function is sufficiently close to the equivocator, then it is sufficiently close to the equivocator regardless of the subset of $\mathbb{E}$ in which it lies, that is, $P \in \Downarrow\mathbb{E}$ implies that if $P \in \mathbb{F} \subseteq \mathbb{E}$, then $P \in \Downarrow\mathbb{F}$, so in particular, $\Downarrow\mathbb{E} \subseteq \Downarrow\Downarrow\mathbb{E}$. This motivates defining $\Downarrow\mathbb{E} \stackrel{df}{=} \mathbb{E}$ in cases where no member of $\mathbb{E}$ is closest to the equivocator and where there are no other considerations that might adjudicate whether or not a function in $\mathbb{E}$ is sufficiently equivocal.

In the remainder of the chapter, we focus on the case in which $\downarrow\mathbb{E}$ is non-empty.

7.3.2 Regularity

There is an important special case. We call the set $\{\mu_1, \ldots, \mu_k\}$ of premise sentences of $\mathcal{L}^\sharp$ *regular* iff it gives rise to a set $\mathbb{P}_{\mu_1,\ldots,\mu_k} = \{P \in \mathbb{P}^{\mathcal{L}} : P \models \mu_1, \ldots, \mu_k\}$ that is closed, convex, and non-empty. Premises are regular, for example, if they are mutually consistent, they involve sentences of $\mathcal{L}$ that are all quantifier-free, and they involve sets of probabilities that are all closed and convex. If the premises are regular, then $\mathbb{E} = \mathbb{P}^* = \mathbb{P}_{\mu_1,\ldots,\mu_k}$ and the set of probabilities to attach to the conclusion sentence is $Y = \{P(\psi) : P \in \Downarrow\mathbb{P}_{\mu_1,\ldots,\mu_k}\}$. This case arises naturally when the premises are observations of empirical probabilities (i.e. $P^*(\varphi_i) = x_i$; the X_i are singletons) or estimates of empirical probabilities (i.e. $P^*(\varphi_i) \in [x_i - \epsilon, x_i + \epsilon]$; the X_i are closed intervals).

Example 7.1 Consider the question $\forall x U x^{3/5} \approx U t_1^?$. Objective Bayesian epistemology interprets the expression on the left-hand side as $P^*(\forall x U x) = 3/5$: this premise is regular, so $\mathbb{E} = \langle \mathbb{P}^* \rangle = \mathbb{P}^* = \{P : P(\forall x U x) = 3/5\}$. There is one function P in $\mathbb{E}$ that is closest to the equivocator, as described in Example 5.7. This function gives $P(Ut_i) = 4/5$, for each constant t_i. Hence, the answer to the question is $4/5$.

7.3.3 Properties of objective Bayesian entailment

Having presented the objective Bayesian semantics, we now investigate some of the properties of the resulting entailment relation.

Definition 7.2. (**Decomposable**) *An entailment relation $\approx$ is called decomposable iff for all sentences $\mu_1, \ldots, \mu_k$ of the language $\mathcal{L}^\sharp$ of the entailment relation, $[\![\mu_1, \ldots, \mu_k]\!] = [\![\mu_1]\!] \cap \cdots \cap [\![\mu_k]\!]$.*

Definition 7.3. (**Monotonic**) *An entailment relation $\approx$ is monotonic iff for all sentences $\nu, \mu_1, \ldots, \mu_k, \ldots, \mu_l$ of the language $\mathcal{L}^\sharp$ of $\approx$ (where $k < l$), we have that $\mu_1, \ldots, \mu_k \approx \nu$ implies $\mu_1, \ldots, \mu_k, \ldots, \mu_l \approx \nu$.*

Proposition 7.4 *A decomposable entailment relation is monotonic.*

Proof: Suppose that $\approx$ is decomposable and that $\mu_1, \ldots, \mu_k \approx \nu$.

Now, $\mu_1, \ldots, \mu_k, \ldots, \mu_l \approx \nu$ iff $P \approx \nu$ for each $P \in [\![\mu_1, \ldots, \mu_l]\!]$. But each such $P \in [\![\mu_1, \ldots, \mu_k]\!]$ since

$$[\![\mu_1, \ldots, \mu_l]\!] = [\![\mu_1]\!] \cap \cdots \cap [\![\mu_l]\!] \subseteq [\![\mu_1]\!] \cap \cdots \cap [\![\mu_k]\!] = [\![\mu_1, \ldots, \mu_k]\!].$$

By assumption, $\mu_1, \ldots, \mu_k \approx \nu$ so for any such P, $P \approx \nu$, as required. ∎

Now, entailment under the objective Bayesian semantics is nonmonotonic: $A_1^{[0,1]} \approx A_1^{.5}$ but it is not the case that $A_1^{[0,1]}, A_1^1 \approx A_1^{.5}$, where A_1 is a propositional variable (respectively atomic sentence) of propositional (respectively predicate) language $\mathcal{L}$. Hence, objective Bayesian entailment is not decomposable.

Although nonmonotonic, this notion of entailment still satisfies a number of interesting and useful properties. In order to characterize these properties, we need to appeal to some notions that are common in the literature on nonmonotonic logics—see, for example, Kraus *et al.* (1990), Makinson (2005, §3.2), and Hawthorne and Makinson (2007) for background. We restrict our attention to the case in which all sets of sentences are regular. A probability function P on $\mathcal{L}$ can be construed as a valuation on $\mathcal{L}^\sharp$: P assigns the value True to sentence μ of $\mathcal{L}^\sharp$ if $P \in \mathbb{P}_\mu$, that is, if $P \models \mu$. Note that $\models$ is a decomposable entailment relation, and, as mentioned in our discussion of the standard semantics, $[\![\mu]\!]_\models = \mathbb{P}_\mu$, for sentence μ of $\mathcal{L}^\sharp$. In particular, if μ is of the form φ^X, then $P \in [\![\mu]\!]_\models$ iff $P(\varphi) \in X$. Recall that for probability functions P and Q, $P \prec Q$ iff P is closer to the equivocator than Q. Now, $(\mathbb{P}^\mathcal{L}, \prec, \models)$ is a *preferential model*: $\mathbb{P}^\mathcal{L}$ is a set of valuations on $\mathcal{L}^\sharp$, $\prec$ is an irreflexive, transitive relation over $\mathbb{P}^\mathcal{L}$, and $\models$ is a decomposable entailment relation. Moreover, since we are focusing on regular sets of sentences, this preferential model is *smooth*: if $P \in [\![\mu]\!]_\models$ then either P is minimal with respect to $\prec$ in $\mathbb{P}_\mu$ or there is a $Q \prec P$ in $\mathbb{P}_\mu$ that is minimal. Hence, this model determines a *preferential consequence relation* $\hspace{0.2em}\mid\hspace{-0.5em}\sim$ as follows: $\mu \hspace{0.2em}\mid\hspace{-0.5em}\sim \nu$ iff P satisfies ν for every $P \in \mathbb{P}^\mathcal{L}$ that is minimal among those probability functions that satisfy μ. Since $\{\mu, \nu\}$ is regular, $\hspace{0.2em}\mid\hspace{-0.5em}\sim$ will agree with $\approx$. Consequently, on regular sets of sentences $\approx$ will satisfy the properties of preferential consequence relations, often called *system-P* properties—see, for example, Kraus *et al.* (1990):

Proposition 7.5. (Properties of Entailment) *Let $\models$ denote entailment in classical logic and let $\equiv$ denote classical logical equivalence. Whenever $\{\mu, \nu, \xi\}$ is regular,*

Right Weakening: *if $\mu \mathrel{|\!\approx} \nu$ and $\nu \models \xi$, then $\mu \mathrel{|\!\approx} \xi$,*
Left Classical Equivalence: *if $\mu \mathrel{|\!\approx} \nu$ and $\mu \equiv \xi$, then $\xi \mathrel{|\!\approx} \nu$,*
Cautious Monotony: *if $\mu \mathrel{|\!\approx} \nu$ and $\mu \mathrel{|\!\approx} \xi$, then $\mu \wedge \xi \mathrel{|\!\approx} \nu$,*
Premise Disjunction: *if $\mu \mathrel{|\!\approx} \nu$ and $\xi \mathrel{|\!\approx} \nu$, then $\mu \vee \xi \mathrel{|\!\approx} \nu$, and*
Conclusion Conjunction: *if $\mu \mathrel{|\!\approx} \nu$ and $\mu \mathrel{|\!\approx} \xi$, then $\mu \mathrel{|\!\approx} \nu \wedge \xi$.*

7.4 A calculus for the objective Bayesian semantics

We now have all the tools we need to say how questions of the form of Equation 7.3, and more generally Equation 7.4, can be answered.

The general strategy is as follows. We interpret a question of the form $\varphi_1^{X_1}, \ldots, \varphi_n^{X_n} \mathrel{|\!\approx} \psi^?$ by means of the objective Bayesian semantics of §7.3. As we saw, this gives an answer in principle to the above question: the objective Bayesian semantics attaches the set $Y = \{P_{\mathcal{E}}(\psi) : P_{\mathcal{E}} \in \Downarrow\langle \mathbb{P}^* \rangle\}$ to the conclusion sentence ψ, where $\mathbb{P}^* = \{P \in \mathbb{P}_{\mathcal{L}} : P \models \chi, \chi \in \biguplus\{\varphi_1^{X_1}, \ldots, \varphi_k^{X_k}\}\}$. The question remains as to how to determine Y in practice. We answer this question by first representing $\mathbb{E} = \Downarrow\langle \mathbb{P}^* \rangle$ by an objective credal net, and then by using this net to calculate the probability interval Y that attaches to ψ.

Example 7.6 Suppose we have a question of the form

$$A_1 \wedge \neg A_2^{[0.8, 0.9]}, (\neg A_4 \vee A_3) \to A_2^{0.2}, A_5 \vee A_3^{[0.3, 0.6]}, A_4^{0.7} \mathrel{|\!\approx} A_5 \to A_1^?$$

This is short for the following question: given that $A_1 \wedge \neg A_2$ has probability between 0.8, and 0.9 inclusive, $(\neg A_4 \vee A_3) \to A_2$ has probability 0.2, $A_5 \vee A_3$ has probability in $[0.3, 0.6]$, and A_4 has probability 0.7, what probability should $A_5 \to A_1$ have? As explained in §7.3, this question can be given an objective Bayesian interpretation: supposing the agent's evidence says $P^*(A_1 \wedge \neg A_2) \in [0.8, 0.9], P^*((\neg A_4 \vee A_3) \to A_2) = 0.2, P^*(A_5 \vee A_3) \in [0.3, 0.6], P^*(A_4) = 0.7$, how strongly should she believe $A_5 \to A_1$? Now, an objective credal net can be constructed to answer this question (in fact, the net is an objective Bayesian net (obnet): $\mathcal{L}$ is a finite propositional language and $\Downarrow\mathbb{E} = \downarrow\mathbb{E}$ is a singleton). First construct undirected constraint graph (Fig. 6.3) by linking variables that occur in the same constraint. Next, follow Algorithm 6.4 to transform the undirected graph into a directed acyclic graph (dag) satisfying the Markov Condition, such as Fig. 6.1. The third step is to maximize entropy to determine the probability distribution of each variable conditional on its parents in the directed graph. This yields the obnet. Finally, we use the net to calculate the probability of the conclusion:

$$P(A_5 \to A_1) = P(\neg A_5 \wedge A_1) + P(A_5 \wedge A_1) + P(\neg A_5 \wedge \neg A_1)$$

$$= P(A_1) + P(\neg A_5|\neg A_1)(1 - P(A_1)).$$

Thus, we must calculate $P(A_1)$ and $P(\neg A_5|\neg A_1)$ from the net, which can be done using standard Bayesian or credal net inference algorithms.

Here is a simple example of a predicate language question:

Example 7.7 Suppose we have a question of the form

$$\forall x(Vx \to Hx)^{[.6,1]}, \forall x(Hx \to Mx)^{[.75,1]}, Vs^{.8} \models Ms^?$$

Again, an obnet can be constructed to answer this question. There is only one constant symbol s, so we can focus on a finite predicate language $\mathcal{L}_1$. Then, the dag satisfying the Markov Condition is depicted in Fig. 6.4. As we saw in §6.2, the corresponding probabilities are $P(Vs) = 4/5, P(Hs|Vs) = 3/4, P(Hs|\neg Vs) = 1/2, P(Ms|Hs) = 5/6$, and $P(Ms|\neg Hs) = 1/2$; together with the dag these probabilities yield an obnet. Standard inference methods then give us $P(Ms) = 11/15$ as an answer to our question.

The general procedure for determining Y can be represented by the following high-level algorithm:

Algorithm 7.8

Input: A question of the form $\mu_1, \ldots, \mu_k \models \psi^?$, where $\{\mu_1, \ldots, \mu_k\}$ is a regular set of sentences of $\mathcal{L}^\sharp$ and ψ is a sentence of $\mathcal{L}$.

1. If $\mathcal{L}$ is an infinite predicate language, let n be the smallest j such that $\mu_1, \ldots, \mu_k$ are sentences of $\mathcal{L}_j^\sharp$ and ψ is a sentence of $\mathcal{L}_j$. Otherwise, let $\mathcal{L}_n = \mathcal{L}$.
2. Construct the constraint graph $\mathcal{G}$ for constraints $\mu_1, \ldots, \mu_k$ on $\mathcal{L}_n$ (Definition 6.2).
3. Transform this graph into a dag $\mathcal{H}$ by means of Algorithm 6.4.
4. Determine the corresponding conditional probability intervals for an objective credal net representation of $\downarrow \mathbb{E}$:
   (a) Determine the credal net on $\mathcal{H}$ that represents the functions, from those that satisfy the Markov Condition with respect to $\mathcal{H}$, that are in $\mathbb{E}$ (i.e. that satisfy the constraints $\mu_1, \ldots, \mu_k$) (Haenni et al. 2010, algorithm 9.1).
   (b) Use Monte-Carlo methods to narrow the intervals in this credal net to represent the probability functions, from those in the above net, that are closest to the equivocator.
5. Use this credal net to determine the interval Y that attaches to ψ (Algorithm 6.1).

Output: An approximation to Y such that $\mu_1, \ldots, \mu_k \models \psi^Y$.

In step 4(b), one can equivalently search for the probability functions with greatest entropy. Entropy is determined by probabilities specified in the network via the identity

$$H(P) = -\sum_{i=1}^{r_n} \sum_{\omega_i} \left(\prod_{A_j \in Anc_i} P(A_j^{\omega_i}|Par_j^{\omega_i}) \right) \log P(A_i^{\omega_i}|Par_i^{\omega_i}).$$

where the w_i range over the states of $Anc_i \stackrel{\text{df}}{=} \{A_i$ and its ancestors in $\mathcal{H}\}$. Hence, one needs to find the range of values of the parameters $P(A_i|Par_i)$ that maximize entropy.

We see then that, given a question of the form of Equation 7.3 or Equation 7.4, an objective credal net can be constructed to provide an answer. Of course, in the worst-case logical inference and probabilistic inference are each computationally intractable (in the sense that computing answers takes time exponential in n), so their combination in a probabilistic logic is no more tractable. But that is the worst case; if the constraint graph is sparse, the objective credal net calculus offers the potential for efficient inference.

8

Judgement Aggregation

In recent years, formal methods originally developed by computer scientists and logicians to deal with inconsistencies in databases have increasingly been applied to problems in social epistemology. Thus, the formal theory of *belief revision and merging*, developed since the 1980s to help maintain consistency when revising and merging sets of propositions, has been applied by Meyer *et al.* (2001) and Chopra *et al.* (2006) to the problem of aggregating preferences, applied by Gauwin *et al.* (2005) to the problem of judgement deliberation and conciliation, and applied by Pigozzi (2006) to the problem of judgement aggregation.

It is this latter application, to judgement aggregation, that is the focus of this chapter. Section 8.1 introduces the problem of judgement aggregation and some of the difficulties encountered in trying to solve this problem. Section 8.2 introduces the theory of belief revision and merging and Pigozzi's application of this theory to judgement aggregation. Section 8.3 argues that judgements should not be merged directly; rather, one should merge the evidence on which the judgements are based. Given this merged evidence, objective Bayesianism and decision theory can be used to derive an appropriate set of judgements—the resulting judgements should be viewed as the aggregate of the judgements of the original individuals (§8.4).

We see, then, that objective Bayesianism is an important component of a solution to the thorny problem of judgement aggregation.

8.1 Aggregating judgements

In many cases, it is important for a collection of agents with their own individual judgements to come to some agreed set of judgements as a group. The question of how best to do this is the problem of judgement aggregation. (In discussions of judgement aggregation, it is normally assumed that a judgement is a proposition or sentence and that by providing a judgement one *endorses* the relevant proposition.)

TABLE 8.1. An example of the discursive dilemma.

	θ	$\theta \to \varphi$	φ
A	True	True	True
B	True	False	False
C	False	True	False
Majority	True	True	False

Simple majority voting on each of the propositions in an agenda is unsatisfactory as a judgement aggregation procedure for the following reason: while each individual may have a consistent set of judgements, the aggregated set of judgements may be inconsistent. This is known as the *discursive dilemma* or the *doctrinal paradox* (Kornhauser and Sager 1986). Table 8.1 displays a simple example, presented by Dietrich and List (2007). Here three individuals, A, B, C, have consistent sets of judgements (e.g. A judges that $\theta, \theta \to \varphi$, and φ), but majority voting yields an inconsistent set $\{\theta, \theta \to \varphi, \neg\varphi\}$ of aggregated judgements.

Problems with majority voting have led to a quest to find a better judgement aggregation procedure. However, a number of impossibility theorems limit the options available—see, for example, List and Pettit (2002, 2004), Dietrich and List (2007), and Pauly and van Hees (2006). In particular, Dietrich and List (2007) show that if the agenda is sufficiently rich and if (formalizations of) the following conditions hold, then the only aggregation functions are *dictatorships*, that is, those that simply take the judgements of a particular individual as the judgements of the group:

Universal Domain: the domain of the aggregation function is the set of all possible profiles of consistent and complete individual judgement sets,

Collective Rationality: the aggregation function generates consistent and complete collective judgement sets,

Independence: the aggregated judgement on each proposition depends only on individual judgements on that proposition, and

Unanimity: if each individual judges a particular proposition true, then so will the aggregate.

(As Dietrich and List 2007 show, this result generalizes Arrow's celebrated impossibility theorem concerning preference aggregation in social choice theory.)

The question thus arises as to whether any reasonable aggregation function remains. Dictatorship aside, is there any reasonable judgement aggregation rule? If so, which of the above conditions does it violate? We see next that Pigozzi (2006) advocates a judgement aggregation rule that violates Independence. But the arguments of §8.4 suggest that judgement aggregation requires a rule that may violate *all* of these conditions. This rule will appeal to a richer epistemology than that invoked by the standard examples of the judgement aggregation literature; in particular, it will distinguish an agent's judgements and the evidence on which those judgements are based. Arguably, only by considering such evidence can one aggregate judgements properly. Thus, problems like that of Table 8.1 turn out to be underspecified.

8.2 Belief revision and merging

The question of how to revise and merge sets of propositions is addressed by the theory of belief revision and merging. *Belief revision* seeks to say how one should revise a deductively closed set of sentences T in the light of new information, a consistent sentence θ. The theory imposes the following desiderata—known as the AGM postulates, after Alchourrón *et al.* (1985)—for a belief revision operator $\star$:

$\star$0: $T \star \theta$ is consistent,
$\star$1: $T \star \theta$ is deductively closed,
$\star$2: $\theta \in T \star \theta$,
$\star$3: if $T + \theta$, the deductive closure of $T \cup \theta$, is consistent, then $T \star \theta = T + \theta$,
$\star$4: if θ and φ are logically equivalent, then $T \star \theta = T \star \varphi$,
$\star$5: if $T \star \theta + \varphi$ is consistent, then $T \star \theta + \varphi = T \star \theta \wedge \varphi$.

Alchourrón *et al.* (1985) also put forward postulates for a *belief contraction* operator $-$, where $T - \theta$ is the result of taking θ away from deductively closed T. Revision and contraction turn out to be related by the Levi identity $T \star \theta = (T - \theta) + \theta$ and by the Harper identity $T - \theta = (T \star \theta) \cap T$ (for θ not a tautology). The theory has been further extended to give an account of *belief update*: while for belief revision θ and T are understood as referring to the same situation, for belief update T is taken to refer to the past and θ to the present. See Katsuno and Mendelzon (1991) and del Val and Shoham (1994).

A third extension of the theory is of special interest here. *Belief merging* seeks to provide an account of how one should combine sets $T_1, \ldots, T_{k_T}$ of sentences. Konieczny and Pino Pérez (1998) put forward the following postulates for a merging operator Δ acting on a multiset $T = \{T_1, \ldots, T_{k_T}\}$, where each T_i is assumed consistent, where the notation T_i may be used to refer to the set of sentences or the conjunction of its members, and where $T \sqcup U$ is the multiset $\{T_1, \ldots, T_{k_T}, U_1, \ldots, U_{k_U}\}$:

Δ0: ΔT is consistent,
Δ1: if $\bigwedge T$ is consistent, then $\Delta T = \bigwedge T$,
Δ2: if T_i and U_i are logically equivalent for $i = 1, \ldots, k_T = k_U$, then ΔT and ΔU are logically equivalent,
Δ3: if $\bigwedge T \wedge \bigwedge U$ is inconsistent, then $\Delta(T \sqcup U)$ does not logically imply T,
Δ4: $\Delta T \wedge \Delta U$ logically implies $\Delta(T \sqcup U)$,
Δ5: if $\Delta T \wedge \Delta U$ is consistent, then $\Delta(T \sqcup U)$ logically implies $\Delta T \wedge \Delta U$.

Konieczny and Pino Pérez (2002) generalize this framework to provide postulates for a merging operator Δ_ι with *integrity constraints* ι that the merged set has to satisfy. Belief merging is related to belief revision by the identity $T \star \iota = \Delta_\iota\{T\}$, where T is a set of sentences and where merging is subject to a deductive closure condition (Lin and Mendelzon 1998, §5; Konieczny and Pino Pérez 2002, §5).

Pigozzi (2006) applies belief merging to the problem of judgement aggregation. The idea is this: if k agents have propositional judgements $T = \{T_1, \ldots, T_k\}$, then one should take $\Delta_\iota T$ as the aggregate of these judgements, where Δ_ι is a particular belief merging operator motivated by majority voting considerations. In the example of the discursive dilemma of Table 8.1, no integrity constraints ι are required, since postulate $\Delta 0$ ensures that the aggregate set is consistent.

Pigozzi (2006) shows that a judicious choice of integrity constraints avoids any paradoxical outcome in another example of the discursive dilemma. However, in all such cases the aggregation procedure results in several equally optimal merged sets—the selection of one of these sets as the aggregate is hence rather arbitrary. Arbitrariness is a fundamental problem when using belief merging techniques to directly aggregate judgement sets, since there is rarely enough information in the judgement sets themselves to allow one to decide which way to resolve an inconsistency in the majority view. But we see next that arbitrariness can be controlled by merging at the level of evidence rather than at the level of judgement.

8.3 Merging evidence

When devising a method of aggregating judgements, one might have one of two goals in mind. One natural goal is to try to find a *fair* procedure—a procedure that treats all agents equally, where all agents' judgements contribute in the same way in determining the aggregate. Since the recent literature on judgement aggregation has stemmed from work in social choice theory—for example, work on trying to devise fair voting systems—this goal has hitherto been paramount. Unfortunately, the discursive dilemma and the impossibility results suggest that this goal may be unattainable.

Another natural goal is to try to find a judgement aggregation procedure that yields the *right judgements*. Arguably, if this is the primary goal, then it matters little whether different agents' judgements play an equal role in determining the aggregate. Thus, the impossibility results, whose assumptions are motivated by fairness considerations, need not apply. This latter goal will be our goal here.[1]

Consider the following judgement aggregation problem. A patient has received some treatment for breast cancer and two consultants, A and B, need to make the following judgements. First they need to judge whether or not the patient's cancer will recur, R, in order to inform the patient of her prospects. Second, they need to judge whether or not chemotherapy is required, C (if recurrence is unlikely, then aggressive treatments such as chemotherapy, which have harsh side effects, may not be justified). In this kind of example, when aggregating judgements it is far more important that the collective judgements be the right judgements than that they be fair to the individual agents' opinions: the patient's health is more important than the egos of the consultants.

Consultant A has clinical evidence: the tumour has not spread to lymph nodes, $\neg L$. This leads her to believe that the cancer will not recur, $\neg R$. She is

---

[1] Bovens and Rabinowicz (2006) and Hartmann *et al.* (2007) pursue this goal, for instance, assessing standard judgement aggregation procedures from the point of view of correctness. This is a normative goal: the question is how one *should* aggregate judgements to yield the right judgements, not how people actually *do* aggregate judgements. Thus, there is no need for an answer to this question to be psychologically realistic. There is a need for any answer to be computationally feasible, but it will be beyond the scope of this chapter to address computational concerns in any detail.

TABLE 8.2. The epistemic states of A and B.

	Grants	Believes	Judges
A	$\neg L$	$\neg R$	$\neg R, \neg C$
B	H	$\emptyset$	R, C

not so sure about whether chemotherapy is required, but on balance she judges that it is not. Thus, A judges $\neg R, \neg C$.

Consultant B has molecular evidence—the presence of certain hormone receptors in the patient, H. This evidence indicates a less favourable prognosis. B is not convinced enough to say that he *believes* R and C, but since a judgement is required, he judges R, C.

Table 8.2 represents the epistemic states of the two consultants. The epistemological picture here is that the agents have three grades of propositions: evidence, beliefs, and judgements. As usual, the agent's evidence includes all that she *takes for granted* in the context of the task in hand. As well as the results of observation, it includes any theoretical and background assumptions that are not currently open to question. (As discussed in §1.4, the evidence base may equally be called her *epistemic background* or *data*. As to whether the agent is rational to take her evidence for granted will depend on the goals of her current enquiry as well as the provenance of the propositions themselves.) The agent's qualitative beliefs are not taken for granted, but are credible enough to be construed as *accepted*, and hopefully rationally so. As can be seen from the above example, judgements often need to be made when there may be little to decide one way or the other: they are *speculated* propositions—again, one hopes, rationally so. Thus, items of evidence are practically certain, beliefs have high credence, while a judgement may be much more weakly believed.

If we are to apply belief revision and merging to the problem of judgement aggregation, the question arises as to what exactly we merge. It is natural to try to merge the judgements themselves—this is the strategy of Pigozzi (2006), for instance. But then we come up against the problem of arbitrariness. In our example, this amounts to merging $\{\neg R, \neg C\}$ with $\{R, C\}$. This can be done by arbitrarily choosing one agent's judgements over the other, or by taking the disjunction of the judgements, $(\neg R \wedge \neg C) \vee (R \wedge C)$, as the aggregate. This last strategy does not avoid arbitrariness: the aggregated judgements do not contain judgements on the propositions we want, so to make judgements about recurrence and chemotherapy on the basis of the disjunctive aggregate requires arbitrary choice.

Perhaps, the theory of belief revision and merging should be applied at the level of belief—this is ostensibly what the theory concerns, after all. This is a mistake, for two reasons. First, the theory of belief revision and merging is more suited to revising and merging evidence than belief. This can be seen as follows. Belief revision is related to nonmonotonic logic (§7.3): there is a correspondence between revision operators $\star$ and consistency-preserving *rational consequence relations* $\mid\!\sim$ (i.e. $\mid\!\sim$ satisfying the Gabbay–Makinson conditions for nonmonotonic consequence together with the rule $\theta \not\models \neg\tau \Rightarrow \theta \not\mid\!\sim \neg\tau$, for tautology τ) via the equivalence $\theta \mid\!\sim \varphi \Leftrightarrow$ either θ is inconsistent or else $\varphi \in T \star \theta$. In turn, nonmonotonic logic is related to probability: there is a correspondence between non-trivial rational consequence relations $\mid\!\sim$ and ϵ-probability functions P (probability functions whose range is $[0, 1]$ augmented with infinitesimal numbers) via the equivalence $\theta \mid\!\sim \varphi \Leftrightarrow$ either $P(\theta) = 0$ or else $P(\neg\varphi|\theta)$ is infinitesimal or zero (see, e.g. Paris 2003). Thus, there is a correspondence between revision operators $\star$ and the ϵ-probability functions P that only award probability zero to contradictions: $T \star \theta$ consists of those sentences whose probability conditional on θ is infinitesimally close to 1. Similarly, there is a correspondence between merging operators Δ_ι (under a deductive closure condition) and such ϵ-probability functions: $\Delta_\iota\{T\}$ consists of those sentences whose probability conditional on ι is infinitesimally close to 1. Now under a Bayesian account, ϵ-probability is construed as a measure of rational degree of belief. Thus, $T \star \theta$ is the set of sentences that should be believed to degree infinitesimally close to 1, given θ; $\Delta_\iota\{T\}$ is the set of sentences that should be believed to degree infinitesimally close to 1, given ι. Under the Bayesian account, degrees of belief are understood as indicative of betting intentions; an infinitesimal difference between two degrees of belief has no consequence in terms of betting; hence, 'infinitesimally close to 1' corresponds to practical certainty. Thus, $\Delta_\iota\{T\}$ is the set of sentences that should be practical certainties given ι. But then $\Delta_\iota\{T\}$ cannot be interpreted as an agent's qualitative beliefs, since practical certainty is an absurdly high standard for qualitative belief. A sentence might be deemed *believed* simpliciter if it is strongly believed—say to degree 0.8 or 0.9 depending on circumstances—but it is far too stringent to insist that one's beliefs must be practically certain. Rather, it is the agent's evidence that is practically certain (at least in the agent's current operating context). Therefore, belief revision and merging should be used to revise and merge evidence rather than belief.[2]

There is a second reason why merging agents' beliefs is mistaken. In our example, such a merging operation would yield $\neg R$ as the merged belief set. Presumably, then, judging $\neg R$ ought to be reasonable given these merged beliefs. But, in fact, such a judgement may be very unreasonable: it may be that the presence of both the patient's symptoms together makes recurrence quite likely. Merging beliefs will thus ignore any interactions between the pieces of evidence that give rise to the beliefs.

These two considerations motivate merging the agents' *evidence* instead of their beliefs or judgements. On the one hand, this is, on reflection, the correct

---

[2]The distinction between evidence and belief tends to be passed over in this literature: Pigozzi (2006) and many of those working in the area of belief revision and merging use 'knowledge' and 'belief' interchangeably. Note that Kyburg Jr et al. (2007, §6) suggest some modifications to the belief revision framework if it is to used to handle propositions that are *rationally accepted*, rather than known. The view of rational acceptance that they consider is that provided by the theory of evidential probability introduced in §7.2.

domain of application of the theory of belief revision and merging. On the other, one must merge the *reasons* for the agents' beliefs and judgements, rather than the beliefs or judgements themselves, if one is to achieve the goal of making the right judgements. The right judgement is that which, considering all available evidence, is most appropriate given the uses to which the judgement will be put. Only by merging the evidence itself can one consider all available evidence; merging beliefs or judgements directly will ignore interactions amongst items of evidence.

Merging at the level of evidence has a further advantage: it reduces arbitrariness. We often have to make judgements in the face of extensive uncertainty—there may be little to decide between a proposition and its negation, yet we must make a call one way or the other. Different agents are likely to make different calls, yielding mutually inconsistent judgements. If these judgements are to be merged, this leads to arbitrariness on the part of the merging operator. Beliefs are of a higher epistemological grade than judgements: to count as a belief in the qualitative sense, a proposition must be strongly believed in a quantitative sense. Fewer propositions are likely to make this grade, and hence there are likely to be fewer inconsistencies between different agents' beliefs than between different agents' judgements. Thus, merging beliefs involves less arbitrariness than merging judgements. But evidence is of a higher grade yet: an item of evidence is taken for granted and so practically certain. Of course, items of evidence may be false; hence, inconsistencies may arise between different agents' data bases. But such inconsistencies will be considerably rarer than those that arise between belief bases, let alone judgements. Therefore, merging evidence avoids much of the arbitrariness that besets attempts to merge judgements directly.

This section has argued that a merging operator should operate on agents' evidence bases rather than their beliefs or judgements. If merging is to be applied to judgement aggregation, we then require some way of determining a set of judgements from a merged evidence base. These judgements can be viewed as the aggregate. It is this task to which we turn next.

8.4 From merged evidence to judgements

Judgement is essentially a decision problem. For each proposition in an agenda, one must decide whether to endorse that proposition or its negation. One makes the *right* judgements to the extent that the judgements are most appropriate considering the uses that will be made of them. Typically, as in the case of our example, one would like one's judgements to be true, but it is sufficient that they are determined in the right way from one's evidence: a false judgement may be the right judgement if it is most plausible on the basis of the limited evidence available; conversely, a true judgement may be the wrong judgement if it is unlikely given what is granted.

Since judgement is a decision problem, decision theory can be applied to judgement formation. A simple decision-theoretic account might proceed like this. Let $U(\theta|\varphi)$ represent the utility of judging θ given that φ is the case. For

TABLE 8.3. A utility matrix for judging recurrence.

		Judgement	
		R	¬R
Case	R	1	−1
	¬R	−3	1

TABLE 8.4. A utility matrix for judging chemotherapy.

		Judgement	
		C	¬C
Case	R	5	−10
	¬R	−4	1

example. Table 8.3 gives a table of utilities for judging cancer recurrence: the judgement will be used to inform the patient of her prospects, so judging the true outcome has positive utility while judging the false outcome has negative utility and is particularly bad when falsely judging recurrence because it leads to needless anxiety. Then, decide on the judgement that maximizes expected utility, $EU(\theta) = \sum_\varphi U(\theta|\varphi)P(\varphi)$, where $P(\varphi)$ is the probability of φ. In our example, judge recurrence if $EU(R) > EU(\neg R)$, that is, if $P(R) > 2/3$. Judgements may depend on propositions other than the judged proposition. Table 8.4 gives a utility matrix for judging chemotherapy: judging in favour of chemotherapy is a good thing if the patient's cancer would otherwise recur but quite bad if it would not recur; judging against chemotherapy is very bad given recurrence but quite good given non-recurrence. Decide in favour of chemotherapy if $EU(C) > EU(\neg C)$, that is, if $P(R) > 1/4$.

We see then that if we are to apply this decision-theoretic approach to judgement, we need to determine probabilities, for example, $P(R)$. This is where objective Bayesianism can be applied. As we have seen, objective Bayesian probabilities are rational degrees of belief and it is an agent's evidence which largely determines the degrees of belief that she should adopt.

We have, now, all the tools we need for a normative account of judgement aggregation. First, use a merging operator to merge the agents' evidence bases. This merged evidence base can be thought of as the evidence base of a hypothetical agent M. Objective Bayesianism can then be used to determine the degrees of belief that M should adopt, given this merged evidence. Finally, decision theory can be used to determine the judgements that M should make given these degrees of belief. These judgements can be viewed as the aggregate of the individual agents' judgements, although they are a function of the individual agents' evidence bases rather than of their judgement sets. (Note that it is *not* assumed that the original agents determine their own judgements via objective Bayesianism and decision theory. Indeed, the procedure outlined here does not depend on the original agents' own judgement procedures at all. All that is assumed is that

TABLE 8.5. The epistemic states of A, B, and the merged agent M.

	Grants	Believes	Judges
A	$\neg L$	$\neg R$	$\neg R, \neg C$
B	H	$\emptyset$	R, C
M	$H, \neg L$	$R^{0.4}$	$\neg R, C$

these agents have some evidence and that this evidence can be made explicit in order for merging to take place. After merging, it is the hypothetical agent who conforms to the norms of objective Bayesian epistemology and decision theory.)

In our example, the items of evidence $\neg L$ and H are consistent. Hence, by merging postulate $\Delta 1$, their merger is $\{\neg L, H\}$. To make it more interesting, we assume that the two consultants have some common evidence gleaned from hospital data, namely that the frequency of recurrence given no lymph cancer is 0.2, and that the frequency of recurrence given the presence of the relevant hormone receptor is 0.6. Thus, the Calibration norm yields the constraints $P(R|\neg L) = 0.2, P(R|H) = 0.6$. We also suppose that the consultants know that R is not a cause of L or H. (As explained in §6.2, in the presence of such causal evidence, entropy must be maximized sequentially: first maximize entropy to find the probability distribution over L and H, then maximize entropy again to find the probability distribution over R, given the previously determined probabilities of L and H.) Then, maximizing entropy yields $P(R) = P(R|\neg L \wedge H) = 2/5$. Now, $P(R) < 2/3$, so M should judge $\neg R$. Further, $P(R) > 1/4$ so M should judge C. The epistemic state of the merged agent is depicted in Table 8.5; rather than M having qualitative beliefs she has a quantitative belief—she believes R to a degree 0.4. Clearly, the aggregate judgement set $\{\neg R, C\}$ could not be produced by merging the agents' judgements, nor could it be produced by merging their beliefs. To find the right judgements, we need to merge evidence.

8.5 Discussion

We have seen that if we are concerned with finding the right judgements, we should not apply belief merging directly to agents' judgement sets. Instead, we should apply belief merging to the agents' evidence bases and consider a single hypothetical agent with this merged evidence base. We can determine appropriate degrees of belief for this hypothetical agent using objective Bayesian theory and then determine appropriate judgements using decision theory. These judgements can be viewed as the aggregate of the original agents' judgement sets.

This judgement aggregation framework appeals to an epistemology of evidence, rational degree of belief, and rational judgement, rather than the epistemology of evidence, qualitative belief, and judgement outlined in §8.3. It integrates logic (merging evidence), probability theory (determining degrees of belief), and decision theory (determining judgements). While objective Bayesianism maps evidence to degrees of belief and decision theory maps degrees of belief and utilities to judgements, the logical stage—merging—is thus far

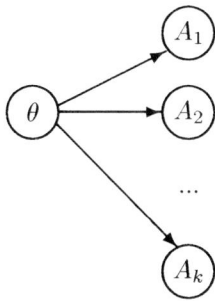

FIG. 8.1. θ causes $A_1, ..., A_k$.

underspecified: no particular merging operator has been advocated for merging evidence. Clearly, a lot will depend on choice of merging operator, and some remarks are in order. Merging operators are often motivated by majority voting considerations: if two agents advocate θ and only one agent advocates $\neg\theta$, then θ is normally taken to be the result of the merger. This strategy is rather dubious when it comes to merging *evidence*, since given that the agents disagree about θ it seems unreasonable to take θ for granted. On the other hand, we do have some information about θ, and it does seem reasonable that the agent should confer higher degree of belief on θ than on $\neg\theta$.[3] One can model this kind of situation as follows. For each agent i who has either θ or $\neg\theta$ in her evidence base, let A_i assert that i grants θ, so that $\neg A_i$ implies that i grants $\neg\theta$. Take θ to be a cause of each A_i, as in Fig. 8.1. Suppose we have some *reliability threshold* $\tau \in [1/2, 1]$ such that for all $i = 1, \ldots, k$, $P(A_i|\theta) \geq \tau$ and $P(\neg A_i|\neg\theta) \geq \tau$. Let A_i^* signify the true literal of A_i (A_i, if i grants θ; $\neg A_i$, if i grants $\neg\theta$) and let m be the *majority* granting θ, that is, the number of agents granting θ minus the number granting $\neg\theta$. Then, we can find the objective Bayesian recommendation by maximizing entropy. The maximum entropy function is represented by the objective Bayesian net (obnet) with the graph of Fig. 8.1 and conditional distributions $P(\theta) = 1/2, P(A_i|\theta) = \tau, P(A_i|\neg\theta) = 1 - \tau$. It yields the following degree of belief in θ on evidence $\mathcal{E} = \{A_1^*, \ldots, A_k^*\}$:

$$P_{\{A_1^*,\ldots,A_k^*\}}(\theta) = \frac{\tau^m}{\tau^m + (1-\tau)^m}.$$

This function is portrayed in Fig. 8.2. We see then that if we have a reliability threshold τ, we have what we need to define a merging operator.

It may seem that the judgement aggregation procedure advocated in this chapter suffers because it appears to ignore the expertise of the two consultants

---

[3]Some merging operators in the literature are not based on majority voting considerations and in this example would not include θ in the merger. In particular, *consistency-based merging operators* look for maximal consistent subsets of the union of the evidence bases—see, for example, Delgrande and Schaub (2007). But these merging operators just ignore the evidence in favour of θ, which itself seems unreasonable.

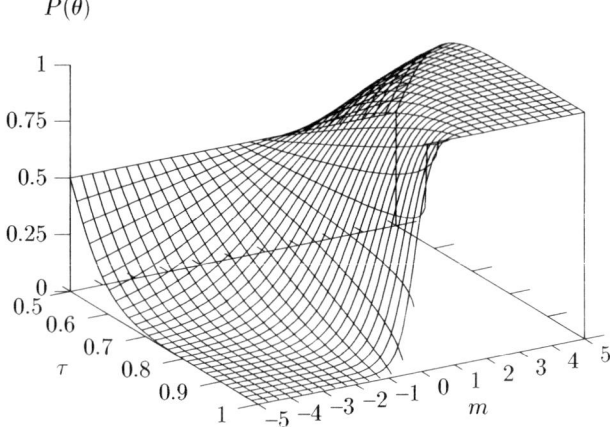

FIG. 8.2. The degree to which the merged agent should believe θ.

in our example: their judgements are aggregated by disregarding their judgements and merging their evidence instead, so the agents themselves appear to play no role in the process. But this latter conclusion does not follow. While this procedure does disregard the agents' actual judgements, their expertise in forming those judgements need not be ignored since it can be put to good use in the formulation of the decision problem. Deciding what to judge on the basis of merged evidence requires the calculation of utilities and here the consultants' expertise will be crucial. Moreover, the aggregation procedure requires that the agents' evidence bases be merged, and this evidence will go beyond a simple list of the patients' symptoms—it will encapsulate a large part of their expertise. Thus, the agents' expertise feeds in to the first and last steps of the aggregation procedure.

There is an interesting question as to how utilities should be determined for the decision-theoretic component of the judgement aggregation procedure. Perhaps, the simplest approach involves aggregating the utilities of the individual agents. Since utilities are numbers rather than propositions, one can simply average the utilities of the respective agents. However, there are several potential problems with this approach. First, the individual agents may have no utilities—they need not have used formal decision theory to come up with their judgements. Second, there may be no guarantee that any requirements of the decision theory (e.g. transitivity of preferences) remain satisfied by the aggregated utilities (Mongin 2001; Hild et al. 2008). Third, just as some agents may be better than others at latching onto the right judgements, some may be better than others at determining appropriate utilities, in which case treating the agents symmetrically may not yield the right utilities. Fourth, the required utilities depend to a certain extent on the uses of the aggregated judgements (e.g. palliative versus curative care of the patient in the cancer example), and these uses may vary considerably from the intentions of the original agents. These considerations motivate an alternative strategy for determining utilities that is analogous to the objective Bayesian procedure for determining degrees of belief. The available evidence, which may include evidence of the original agents' utilities and intentions, imposes constraints on an appropriate utility function for the decision-theoretic component of the aggregation problem (this is the analogue of the Calibration norm of objective Bayesianism). The decision theory will also impose certain constraints (cf. the Probability norm). One may then choose a utility function, from all those that satisfy the constraints, that is as equivocal as possible (Equivocation). Exactly how these principles are to be fleshed out remains a question for future research.

The normative goal sets the approach of this chapter apart from previous work in judgement aggregation, which tends to be motivated by considerations of fairness. In particular, the impossibility results discussed in §8.1 do not get off the ground because the assumptions they make need not hold. Universal Domain fails because the individual judgement sets are not even in the domain of the aggregation function. (One can salvage Universal Domain by construing 'judgements' liberally to include evidence as well as judgements in the literal sense, but there seems little to be gained by this since the whole approach outlined here is predicated upon a distinction between evidence and judgement.) As to whether Collective Rationality holds will depend on the particular decision theory employed; consistency may seem desirable but it is not too hard to envisage very complex agendas where some more limited paraconsistency is all that is required. Independence fails because the aggregate judgements depend on the agents' evidence bases rather than their judgements. Unanimity fails because each agent's evidence may motivate a judgement θ while the collective evidence indicates $\neg\theta$ (as happens with the lottery paradox, for instance). Suppose, for example, that all agents grant the same item of evidence in favour of θ while each agent has her own, slightly weaker, piece of evidence against θ, and that these latter items of evidence are all independent. Then, while in each agent's case the common favourable evidence outweighs the particular agent's unfavourable evidence, the aggregate of the unfavourable evidence may outweigh the aggregate of the favourable evidence (which is just a single item). Note that it is not assumed that each individual agent is rational, so this scenario may also occur when all agents have the same item of evidence against θ and each misreads the evidence, taking it to be evidence in favour of θ.

Returning to the discursive dilemma of Table 8.1, we see that the problem is underspecified. There is no obvious solution to the paradox because there is no information about the evidence on which the judgements are based. Only by considering the total evidence available can we determine the right judgements.

9

Languages and Relativity

Having developed objective Bayesianism on propositional and predicate languages, we turn, in §9.1, to a richer language, the language of the mathematical theory of probability. We see that while we gain in expressibility we lose in terms of logical structure and that this can make the Equivocation norm harder to apply. Section 9.2 responds to the objection that objective Bayesian probability suffers from a pathological relativity to the agent's language $\mathcal{L}$: it is by no means clear that such relativity as there is, is pathological. Similarly, we see in §9.3 that although objective Bayesianism is open to the charge of *subjectivity* in some cases, it is no pathological kind of subjectivity. Moreover, the relativity of objective Bayesian probability to an agent's evidence and language can in principle be eliminated, leading to an *ultimate belief* notion of probability.

9.1 Richer languages

So far, in this book we have considered finite propositional languages and simple first-order predicate languages. But, it is possible to define probability over much richer languages. Thus, Gaifman and Snir (1982) consider first-order predicate languages of arithmetic and show that they can be used as the basis for a certain amount of Bayesian statistics; Scott and Kraus (1966) consider infinitary languages—that is, first-order languages that admit infinite conjunctions and disjunctions. Following Howson (2009, §§2–3), one might wonder whether, given the quirks and complexities of such richer languages, it might not be better simply to progress to the usual mathematical language for probability theory, that of fields of sets and of random variables. We explore this approach in §9.1.1.

9.1.1 *The language of probability theory*

A function P from a field $\mathcal{F}$ of subsets of an outcome space Ω to the non-negative real numbers is a probability function just if it satisfies Countable Additivity (§2.1). Now, Ω typically represents *single-case* outcomes: for example, when considering a particular roll of a particular die one might have $\Omega = \{1, 2, 3, 4, 5, 6\}$ and $\mathcal{F} = \{\emptyset, \{1, 3, 5\}, \{2, 4, 6\}, \Omega\}$; when considering two particular rolls of the die one might have $\Omega' = \Omega \times \Omega$, $\mathcal{F}' = \mathcal{F} \times \mathcal{F}$, and so on. For a single roll and each $\omega \in \Omega$, ω can be taken to express the proposition that the roll of the die will yield ω spots uppermost, and for each $F \in \mathcal{F}$, F can be taken to express the proposition that the roll will be such that the number of spots uppermost will lie in F. The $\omega \in \Omega$, then, can be construed as elementary sentences, analogous to the atomic states ω considered previously. Moreover, the $F \in \mathcal{F}$ can be construed as compound sentences with set union and complement corresponding to disjunction and negation, respectively. Since P is defined over $\mathcal{F}$, $\mathcal{F}$ can be taken as the set $S\mathcal{L}$ of sentences of an agent's language. (Note that no two such sentences can be logically equivalent—there is at most one way to express any given proposition; this is not a problem in our context since in any case the Probability norm awards logically equivalent expressions the same probability.) Here, then, $\mathcal{L}$ refers to the language of the mathematical theory of probability, whose sentences are in $\mathcal{F}$. If Ω is countable and each $\omega \in \mathcal{F}$ (in which case $\mathcal{F}$ is the set of all subsets of Ω), then the probability function P is determined by the probabilities of the elementary sentences ω, as is the case in the languages considered hitherto; hence, this formulation is a natural generalization of the formalism of earlier chapters.

The language of the mathematical theory of probability is enhanced by the use of random variables. A *simple random variable* is a function X from Ω to a finite set of real numbers such that $\{\omega : X(\omega) = x\} \in \mathcal{F}$ for each real number x. One might, for example, define a simple random variable X by $X(\omega) = 1$ if the roll of the die yields an even outcome and 0 if it yields an odd outcome; the statement $X = 1$ then corresponds to the sentence $\{\omega : X(\omega) = 1\} = \{2, 4, 6\}$ and the probability of the former statement is identified with that of the latter sentence. More generally, a *random variable* is a function X from Ω to the real numbers such that $\{\omega : X(\omega) \in H\} \in \mathcal{F}$, for each linear Borel set H (the linear Borel sets are the members of the σ-field generated by the intervals $(a, b]$ for real numbers a, b; a σ-field is a field that is closed under countable unions; the σ-field generated by the intervals is the intersection of all the σ-fields containing the intervals; see Billingsley (1979) for a fuller discussion). A statement of the form $X \in H$ is translated as $\{\omega : X(\omega) \in H\}$ which is in $\mathcal{F}$, so its probability is determined by P. More complex statements, such as those of the form $\lim_{n \to \infty}(X_1 + \cdots + X_n)/n \in H$ can then be assigned probabilities derivatively. In sum, while the use of random variables yields a rich language, all questions of interest are determined by the probability space $(\Omega, \mathcal{F}, P)$ and it will suffice for us to focus our attention on this space.

9.1.2 *Norms*

Having settled on a language, we now turn to the norms of objective Bayesianism. The Probability and Calibration norms are straightforward:

Probability: The strengths of an agent's beliefs should be representable by a probability function.

Here, a probability function P on $\mathcal{L}$ is a function from the set $S\mathcal{L} = \mathcal{F}$ of sentences of $\mathcal{L}$ to the non-negative real numbers that satisfies Countable Additivity (§2.1). This version of the Probability norm can be justified by Dutch book considerations similar to those in §§3.2, 5.2 (Williamson 1999, §5):

Theorem 9.1 Define function $P : S\mathcal{L} \longrightarrow \mathbb{R}$ by $P(F) =$ the agent's betting quotient for F. The agent's bets on propositions expressible in $\mathcal{L}$ avoid the possibility of a Dutch book iff $P(F) \geq 0$, for each $F \in S\mathcal{L}$, and:

Countable Additivity: If $E_1, E_2, \ldots \in \mathcal{F}$ partition Ω, then $\sum_{i=1}^{\infty} P(E_i) = 1$.

Proof: First, we show that a Dutch book is avoided only if the conditions are satisfied.

That $P(F) \geq 0$ follows as for P1 in Theorem 3.2. Similarly, one can show that $P(\Omega)$ must be 1—see the proof of P2 in Theorem 3.2.

If Countable Additivity fails, then a Dutch book is possible, as we now see. If $\sum_{i=1}^{\infty} P(E_i) > 1$, then to ensure finiteness choose some k such that $\sum_{i=1}^{k} P(E_i) > 1$. Set stakes

$$S_F = \begin{cases} 1 & : \quad F = E_i, i = 1, \ldots, k \\ -1 & : \quad F = \Omega \\ 0 & : \quad \text{otherwise} \end{cases}.$$

The agent's total loss is then

$$L = \sum_{i=1}^{k} P(E_i) - \sum_{i=1}^{k} I_{E_i} - P(\Omega) + I_\Omega.$$

But, this is positive whatever happens because $I_\Omega - \sum_{i=1}^{k} I_{E_i} \geq 0$ and $P(\Omega) = 1$, so we have a Dutch book.

On the other hand, if $\sum_{i=1}^{\infty} P(E_i) < 1$ set

$$S_F = \begin{cases} -1 & : \quad F = E_i, i = 1, 2, \ldots \\ 1 & : \quad F = \Omega \\ 0 & : \quad \text{otherwise} \end{cases}.$$

The agent's total loss is then

$$L = -\sum_{i=1}^{\infty} P(E_i) + \sum_{i=1}^{\infty} I_{E_i} + P(\Omega) - I_\Omega = 1 - \sum_{i=1}^{\infty} P(E_i) > 0,$$

a Dutch book.

Conversely, if the conditions hold then no Dutch book is possible. The loss incurred on betting on the partition when E_i is true is

$$L_i = \sum_{j=1}^{\infty} (P(E_j) - I_{E_j}) S_{E_j} = \sum_{j=1}^{\infty} P(E_j) S_{E_j} - S_{E_i}.$$

Under our constraint that only a finite amount of money changes hands, the sum $\sum_{j=1}^{\infty} P(E_j) S_{E_j}$ must be finite. Hence, the expected loss is

$$\sum_{i=1}^{\infty} P(E_i) L_i = \sum_{i=1}^{\infty} P(E_i) \left[\sum_{j=1}^{\infty} P(E_j) S_{E_j} - S_{E_i} \right]$$

$$= \sum_{i=1}^{\infty} P(E_i) \sum_{j=1}^{\infty} P(E_j) S_{E_j} - \sum_{i=1}^{\infty} P(E_i) S_{E_i}$$

$$= 1 \sum_{j=1}^{\infty} P(E_j) S_{E_j} - \sum_{i=1}^{\infty} P(E_i) S_{E_i}$$

$$= 0$$

in which case it cannot be that $L_i > 0$ for all i, as required. ∎

The Calibration norm remains unchanged:

Calibration: The agent's degrees of belief should satisfy constraints imposed by her evidence.

Here, as before, we take $\mathbb{E} = \langle \mathbb{P}^* \rangle \cap \mathbb{S}$ to be the set of probability functions that are compatible with the agent's evidence $\mathcal{E}$, and explicate the Calibration norm as saying that the agent's belief function $P_{\mathcal{E}}$ should be in $\mathbb{E}$.

Equivocation: The agent's degrees of belief should otherwise be sufficiently equivocal.

In explicating the Equivocation norm the two main tasks are of course to say what the equivocator is and to say how to measure closeness to the equivocator. But, on the language of the mathematical theory of probability, these tasks are not so straightforward. We examine these tasks in turn.

9.1.3 *Potential equivocators*

While on a propositional or predicate language, it is intuitively clear that the equivocator should and can equivocate between the atomic states, the language of the mathematical theory probability—which increases expressibility at the expense of logical structure—does not on its own determine a natural equivocator. One might suggest that the equivocator should equivocate between the elementary outcomes $\omega \in \Omega$. But, these outcomes may not be in the domain $\mathcal{F}$ of the probability function, that is, they may not be expressible in the agent's language. Even if they are in the agent's language, equivocating between them may not fully determine a probability function. Worse, there may be *no* probability function on $\mathcal{F}$ that equivocates between the elementary outcomes: if Ω is the unit interval, say, and $\mathcal{F}$ is the set of its subsets, then there is no probability function P on $\mathcal{F}$ that sets $P(\{\omega\}) = 0$, for all elements ω of the unit interval (Billingsley 1979, p. 46).[1] Therefore, more information is required before one can pin down a suitable equivocator, and indeed before one can be sure that some such equivocator exists.

Often this extra information is implicit in the agent's current operating context. The agent may be trying to solve a particular problem, or may be regularly

---

[1] This result rests on the continuum hypothesis as well as the axiom of choice, and so may be open to debate.

solving a particular kind of problem, and the problem specification itself can help determine an appropriate equivocator. Thus, Jaynes (2003, chapter 12) argues that a problem specification can often determine certain symmetries that should be satisfied. If one can transform a problem in a certain way that yields essentially the same problem, then arguably the equivocator for the transformed problem should be the same as that for the original problem. Thus, the equivocator should be invariant under the corresponding transformations of its parameters. Jaynes (1973, 2003, §12.4.4) argues that this method gets a round some of the supposed paradoxes of the principle of indifference, for example solving Bertrand's paradox by successfully answering the question: how strongly should you believe that a long straw tossed randomly onto a circle falls in such a way that the chord that the straw defines is longer than the side of the inscribed equilateral triangle? Similarly, Mikkelson (2004) argues that this method solves the wine–water paradox by successfully answering the question: if wine is mixed with water and there is at most three times as much of one as the other, how strongly should you believe that there is no more than twice as much wine as water? However, in this case the invariance requirement does not uniquely isolate Mikkelson's solution (Deakin 2006). In general, this kind of invariance argument yields one or more *constraints* that the equivocator should satisfy; it need not uniquely determine the equivocator. So the objective Bayesian method then becomes: set your degrees of belief according to a function in $\mathbb{E}$ that is sufficiently close to some function in $\mathbb{P}_=$, the set of *potential equivocators*, that is, those functions that satisfy the relevant invariance constraints. Letting $d(P, \mathbb{P}_=) = \inf\{d(P, Q) : Q \in \mathbb{P}_=\}$, the advice is to choose a function in $\mathbb{E}$ sufficiently close to $\mathbb{P}_=$ in the sense of this distance function.

Objective Bayesian statisticians have developed several other methods for determining an appropriate equivocator (or *reference prior* as it is often called)—see Kass and Wasserman (1996) for a survey. It should be noted that a potential equivocator $P_=$ does not always turn out to be a probability function: as we saw above, there may be no probability function on $\mathcal{F}$ which satisfies the required invariance constraints. The reference prior is then sometimes called an *improper prior*. This is problematic when Bayesian conditionalization is used as an updating method: updating the equivocator on evidence $\varepsilon \in \mathcal{F}$ yields a belief function $P_=(\cdot|\varepsilon)$ that may not be a probability function, in violation of the Probability norm. But there is no such problem under the framework advocated in this book: whether or not potential equivocators are probability functions, $\mathbb{E}$ is a set of probability functions and hence so is $\Downarrow\mathbb{E}$, the set of those functions in $\mathbb{E}$ that are sufficiently close to some potential equivocator.

Arguably, the role of the infinite is just to facilitate our reasoning about the large but finite and discrete universe in which we dwell (Hilbert 1925). In which case, the chief constraint on the behaviour of the infinite is that it conserves the behaviour of the finite. Thus, in Chapter 5 it was important that in defining objective Bayesianism on an infinite predicate language it should behave very much like objective Bayesianism on a large but finite predicate language.

The same is so with continuous languages, for example, a language whose sentences are represented by a field $\mathcal{F}$ of subsets of the real line. Suppose, then, that $\mathcal{L}_n = \{A_1, \ldots, A_{r_n}\}$ is a finite propositional or predicate language and that $\mathcal{L}$ is a language, of the form considered in this section, that extends $\mathcal{L}_n$ in the sense that each A_i can be translated as some $A_i^{\mathcal{L}} = F_i \in \mathcal{F}$. If an agent's native language is $\mathcal{L}_n$, she might be tempted to switch to $\mathcal{L}$ to draw conclusions that could be drawn—but drawn with more difficulty—using $\mathcal{L}_n$. It is important then that $\mathcal{L}$ be a conservative extension of $\mathcal{L}_n$ in the sense that $P_{\mathcal{E}}^{\mathcal{L}} \lvert_{\mathcal{L}_n^{\mathcal{L}}}$ should match $P_{\mathcal{E}}^{\mathcal{L}_n}$, where $\mathcal{L}_n^{\mathcal{L}}$ is the sublanguage of $\mathcal{L}$ whose sentences are represented by the field generated by $\{A_1^{\mathcal{L}}, \ldots, A_{r_n}^{\mathcal{L}}\}$. In particular, this yields constraints that should be satisfied by any potential equivocator on $\mathcal{L}$: $P_{=}^{\mathcal{L}}(\omega_n^{\mathcal{L}}) = P_{=}^{\mathcal{L}_n}(\omega_n) = 1/2^{r_n}$, where $\omega_n^{\mathcal{L}}$ is the translation of atomic state $\omega_n \in \Omega_n$ into a sentence of $\mathcal{L}$.

In sum, a set $\mathbb{P}_=$ of potential equivocators can be determined by appealing to constraints imposed by invariance considerations, constraints imposed by considerations to do with conservative extension, and other objective Bayesian statistical techniques.

9.1.4 *Closeness*

We now turn to the second task: that of saying what it is to be close to a potential equivocator. As discussed in §3.4, the notion of distance from the equivocator is tied to the loss function in play in the current operating context. As before, though, we will restrict our attention to logarithmic loss and cross-entropy divergence.

If Ω is finite, then there is a finite finest partition of Ω by $\mathcal{F}$-sets, say $\{F_1, \ldots, F_{r_n}\}$. (Of course, if $\{\omega\} \in \mathcal{F}$ for each $\omega \in \Omega$, then these sets yield the finest partition.) Then, the divergence measure of §2.3 can be applied here:

$$d(P, Q) = \sum_{i=1}^{r_n} P(F_i) \log \frac{P(F_i)}{Q(F_i)}.$$

Plausibly, in this case $P_=(F_i) = 1/2^{r_n}$ for each i, and we proceed as in the case of a propositional language or finite predicate language.

If Ω is infinite and there is a countably infinite finest partition of Ω by $\mathcal{F}$-sets with a natural ordering, say $\{F_1, F_2, \ldots\}$, then one can take limits

$$d(P, Q) = \sum_{i=1}^{\infty} P(F_i) \log \frac{P(F_i)}{Q(F_i)}.$$

In this case, though, it is hard to conceive of a natural equivocator that is a probability function. Plausibly, perhaps, $\mathbb{P}_= = \{P : P(F_i) = x$ for all i and some real $x\}$, in which case $\mathbb{E}$ may well admit no function closest to a potential equivocator (indeed, divergence from a potential equivocator may be negative). As usual, as to which functions are *sufficiently* close to $\mathbb{P}_=$ will depend on pragmatic considerations.

If Ω is infinite and there is a natural sequence of countable partitions of Ω by $\mathcal{F}$-sets, each partition in the sequence being a refinement of the previous, then we can proceed as we did in Chapter 5. Here for the nth partition in the sequence

$$d_n(P,Q) = \sum_i P(F_i) \log \frac{P(F_i)}{Q(F_i)},$$

and P is closer to Q than R if $d_n(P,Q) < d_n(R,Q)$ for sufficiently large n.

In the continuous case, if P, Q, and R are zero on the same sentences of $\mathcal{F}$, then there are non-negative functions p, q on Ω such that $P(F) = \int_F p\,dR$ and $Q(F) = \int_F q\,dR$, for all $F \in \mathcal{F}$; this is a consequence of the Radon–Nikodym theorem (Billingsley 1979, Theorem 32.2) and p and q are called *Radon–Nikodym derivatives* of P and Q with respect to R. As Kullback and Leibler (1951, §2) show, we can define

$$d(P,Q) = \int_\Omega p \log \frac{p}{q}\,dR.$$

This divergence function can act as a suitable measure of closeness to a potential equivocator.

By way of example, let us return to the wine–water paradox. Here the evidence $\mathcal{E}$ is that a gallon of liquid is composed of wine and water and there is at most three times as much of one as of the other. Now an agent whose finite language can express statements about various quantities of wine and statements about various quantities of water might be content to move to a language based on $\Omega = [0,1] \times [0,1]$, where the first element of the pair denotes the volume of wine in gallons and the second denotes the volume of water in gallons, with sentences $\mathcal{F} = \mathcal{B} \times \mathcal{B}$, where $\mathcal{B}$ is the σ-field of Borel sets (generated by subintervals of the unit interval). Given that the equivocator on a finite language would equivocate between the different quantities of wine and equivocate between the various quantities of water, for various units, the natural equivocator on the richer language is $P_= = L \times L$, where L is the uniform probability function defined by $L([a,b]) = b - a$. Define random variables $X(x,y) = x$, the volume of wine; $Y(x,y) = y$, the volume of water; $Z_1 = X/Y$, the ratio of wine to water; and $Z_2 = Y/X$, the ratio of water to wine. The evidence can then be expressed as $\mathcal{E} = \{X+Y = 1, 1/3 \leq Z_i \leq 3\}$. The set of probability functions compatible with this evidence is $\mathbb{E} = \{P : P([1/4, 3/4] \times [1/4, 3/4]) = 1, P(X+Y=1) = 1\}$. The function $P_\mathcal{E}$ closest to the equivocator is that which is uniform over the volume of wine (respectively water) in $[1/4, 3/4]$. If we ask how strongly the agent should believe that there is no more than twice as much wine as water, the answer is $P(Z_1 \leq 2) = 5/6$, in accordance with the proposal of Mikkelson (2004).

In cases other than those outlined above, we cannot speak of closeness *simpliciter*, but rather, closeness in a particular *respect*. A *respect* can be explicated by a random variable. A random variable X and probability function P have density p with respect to Lebesgue measure if p is a non-negative Borel function on the real numbers (applying f^{-1} to a linear Borel set yields a linear Borel set) and $P(X \in H) = \int_H p(x)\,dx$ for each linear Borel set H. Given a respect X, we can define the X-*divergence* of P from Q to be

$$d_X(P,Q) = \int_{-\infty}^{\infty} p(x) \log \frac{p(x)}{q(x)}\,dx,$$

where p is the density of (X, P) and q is the density of (X, Q). P is then closer to Q than R in respect to X if $d_X(P,Q) < d_X(R,Q)$. One can go on to construe the set $\Downarrow\!\mathbb{E}$ of functions in $\mathbb{E}$ that are sufficiently close to $\mathbb{P}_=$ to be those that are sufficiently close in the relevant respects, where the relevant respects are determined by contextual considerations such as problem formulation.

We see, then, that the language of the mathematical theory of probability is expressive, but because it abstracts away from logical considerations it offers little in the way of guidance concerning the implementation of the Equivocation norm. One needs to appeal to other considerations in order to determine an appropriate set $\mathbb{P}_=$ of potential equivocators and to determine respects in which closeness to these potential equivocators is desired.

9.2 Language relativity

Objective Bayesianism has been criticized for being language- or representation-dependent: it has been argued that the Maximum Entropy Principle awards the same event different probabilities depending on the way in which the problem domain is formulated.

John Maynard Keynes surveyed several purported examples of language dependence in his discussion of Laplace's Principle of Indifference (Keynes 1921). As we saw in §2.2, this latter principle advocates assigning the same probability to each of a number of possible outcomes in the absence of any evidence which favours one outcome over the others. Keynes (1921, §4.21) added the condition that the possible outcomes must be indivisible. The Maximum Entropy Principle makes the same recommendation in the presence of evidence that is symmetric with respect to the indivisible outcomes and so inherits any language dependence of the Principle of Indifference.

A typical example of language dependence proceeds as follows (Halpern and Koller 1995, §1). Suppose an agent's language can be represented by the propositional language $\mathcal{L} = \{C\}$ with just one propositional variable C, which asserts that a particular book is colourful. The agent has no evidence, and so by the Principle of Indifference (or equally by the Maximum Entropy Principle) assigns $P(C) = P(\neg C) = 1/2$. But now consider a second language $\mathcal{L}' = \{R, B, G\}$, where R signifies that the book is red, B that it is blue, and G that it is green. An agent with no evidence will give $P(\pm R \wedge \pm B \wedge \pm G) = 1/8$. Now $\neg C$ is equivalent to $\neg R \wedge \neg B \wedge \neg G$, yet the former is given probability $\frac{1}{2}$ while the latter is given probability $\frac{1}{8}$. Thus, the probability assignments of the Principle of Indifference and the Maximum Entropy Principle depend on the choice of language.

Paris and Vencovská (1997) offer the following resolution. They argue that the Maximum Entropy Principle has been misapplied in this type of example: if an agent refines the propositional variable C into $R \vee B \vee G$, one should consider not $\mathcal{L}'$ but $\mathcal{L}'' = \{C, R, B, G\}$ and make the agent's evidence, namely $C \leftrightarrow R \vee B \vee G$, explicit. If we do that, then the probability function on $\mathcal{L}''$ with maximum entropy, out of all those that satisfy the evidence—that is, which assign $P(C \leftrightarrow R \vee B \vee G) = 1$—will yield a value $P(\neg C) = 1/2$. This is just the same value as that given by the Maximum Entropy Principle on $\mathcal{L}$ with no evidence. Thus, there is no inconsistency.

This resolution is all well and good if we are concerned with a single agent who refines her language. But the original problem may be construed rather differently. If *two* agents have languages $\mathcal{L}$ and $\mathcal{L}'$, respectively, and no evidence, then they assign two different probabilities to what we know (but they do not know) is the same proposition. There is no getting round it: probabilities generated by the Maximum Entropy Principle depend on language as well as evidence.

Interestingly, language dependence in this latter multilateral sense is not confined to the Maximum Entropy Principle. As Halpern and Koller (1995) and Paris and Vencovská (1997) point out, there is no non-trivial principle for selecting rational degrees of belief which is independent of language in the multilateral sense. More precisely, suppose we want a principle that selects a set $\mathbb{O}_\mathcal{E}$ of probability functions that are optimally rational on the basis of an agent's evidence $\mathcal{E}$. If $\mathbb{O}_\mathcal{E} \subseteq \mathbb{E}$, that is, if every optimally rational probability function must satisfy constraints imposed by $\mathcal{E}$, and if $\mathbb{O}_\mathcal{E}$ ignores irrelevant information inasmuch as $\mathbb{O}_{\mathcal{E} \cup \mathcal{E}'}(\theta) = \mathbb{O}_\mathcal{E}(\theta)$ whenever $\mathcal{E}'$ involves no propositional variables in sentence θ, then the only candidate for $\mathbb{O}_\mathcal{E}$ that is multilaterally language-independent is $\mathbb{O}_\mathcal{E} = \mathbb{E}$ (Halpern and Koller 1995, Theorem 3.10). Only empirically based subjective Bayesianism is multilaterally language-independent.

So much the better for empirically based subjective Bayesianism and so much the worse for objective Bayesianism, one might think. But such an inference is too quick. It takes the desirability of multilateral language independence for granted. Arguably, however, language independence is not as compelling a desideratum as might seem at first sight. This is because we expect an agent's rational degrees of belief to vary with her evidence, but there is a sense in which an agent's language constitutes empirical evidence. Language evolves to better latch onto the world. Hence, that an agent adopts a particular language $\mathcal{L}$ rather than an arbitrary language $\mathcal{L}'$ is evidence that the important and natural properties in the agent's environment are better captured by predicates in $\mathcal{L}$ than by those in $\mathcal{L}'$; that predicates of $\mathcal{L}$ are more likely to be relevant to each other than those in $\mathcal{L}'$: that $\mathcal{L}$ offers a more efficient coding of the messages that the agent is likely to pass, and perhaps even that the atomic states of $\mathcal{L}$ are more suitable than those of $\mathcal{L}'$ as a partition over which to equivocate (Williamson 2005a, chapter 12). For example, having dozens of words for snow in one's language says something about the environment in which one lives; if one is going to equivocate about the weather tomorrow, it is better to equivocate between the basic states definable in one's own language than in some arbitrary other language. Granted that language itself is a kind of evidence, and granted that an agent's degrees of belief should depend on her evidence, language independence becomes a rather dubious desideratum.

Note that while Howson (2001, p. 139) criticizes the Principle of Indifference on account of its language dependence, the example he cites can be used to support the case *against* language independence as a desideratum. Howson considers two first-order languages with equality: $\mathcal{L}_1$ has just a unary predicate U while $\mathcal{L}_2$ has unary U together with two constants t_1 and t_2. The explicit evidence $\mathcal{E}$ is just 'there are exactly 2 individuals', while sentence θ is 'something has the property U'. $\mathcal{L}_1$ has three models of $\mathcal{E}$, which contain 0, 1, and 2 instances of U, respectively, so $P(\theta) = 2/3$. In $\mathcal{L}_2$ individuals can be distinguished by constants, and thus there are eight models of $\mathcal{E}$ (if constants can name the same individual), six of which satisfy θ, so $P(\theta) = 3/4 \neq 2/3$. While this is a good example of language dependence, the question remains whether language dependence is a problem here. As Howson himself hints, $\mathcal{L}_1$ might be an appropriate language for talking about bosons, which are indistinguishable, while $\mathcal{L}_2$ is more suited to talk about classical particles, which are distinguishable, and thus able to be named by constants. Hence, choice of language $\mathcal{L}_2$ over $\mathcal{L}_1$ indicates distinguishability, while conversely choice of $\mathcal{L}_1$ over $\mathcal{L}_2$ indicates indistinguishability. In this example, then, language betokens implicit evidence. Of course, all but the most ardent subjectivists agree that an agent's degrees of belief ought to be influenced by her evidence. Therefore, language independence becomes an inappropriate desideratum.

In sum, while the Principle of Indifference and the Maximum Entropy Principle have both been dismissed on the grounds of language dependence, it seems clear that some dependence on language is to be expected if degrees of belief are to adequately reflect implicit as well as explicit evidence. So much the better for objective Bayesianism, and so much the worse for empirically based subjective probability which is language-invariant.

9.3 Objectivity

Just as subjective Bayesianism is not entirely subjective—it holds that you should not adopt degrees of belief that are non-probabilistic for instance—similarly objective Bayesianism falls short of total objectivity. There are two senses in which objective Bayesianism admits a certain amount of subjectivity or relativity. First, an agent's evidence and language may not uniquely determine an appropriate belief function, in which case there is some room for subjective choice as to how strongly to believe certain propositions. Second, under an objective Bayesian interpretation, probability is relative to evidence and language in the first place. We consider these two forms of subjectivity in turn.

9.3.1 Non-uniqueness

An agent's evidence and language may not uniquely determine a belief function: the set of sufficiently equivocal evidentially compatible probability functions, $\Downarrow\mathbb{E}$, may contain more than one function. Since an agent is rationally entitled to set her degrees of belief according to any one of these belief functions, objective Bayesianism permits an element of subjective choice. This choice is one of the features that distinguishes objective Bayesianism from the logical interpretation of probability, which requires uniqueness wherever it is defined (§2.2).

We have seen that this non-uniqueness can arise in a variety of ways. For instance, $\mathbb{E}$ may not be closed (§2.3); similarly, on a predicate language it may admit infinite descending chains with respect to the closeness relation (§5.4); or the language may not determine a fixed ordering of the constants (§5.6); or $\downarrow\mathbb{E}$ may contain multiple functions equidistant from the equivocator (§5.7); or the language may lack the structure to determine a single equivocator (§9.1); or there may be no single most appropriate closeness relation (§§3.4, 9.1). Of course, even in cases in which there is a single maximally equivocal belief function, there may be multiple functions that are sufficiently equivocal, just by dint of tolerance of numerical inaccuracies, for example (§2.3). The most we can say in general is that $\emptyset \subseteq \downarrow\mathbb{E} \subseteq \Downarrow\mathbb{E} \subseteq \mathbb{E}$ and $\emptyset \neq \Downarrow\mathbb{E}$. In particular, objective Bayesianism is no more subjective than empirically based subjective Bayesianism.

Is such subjectivity a problem? Perhaps not. Although objective Bayesianism often yields objectivity, it can hardly be blamed where none is to be found. If there is nothing to decide between two belief functions, then subjectivity simply does not matter. Under such a view, all the Bayesian positions—strict subjectivism, empirically based subjective probability and objective Bayesianism—accept the fact that selection of degrees of belief can be a matter of arbitrary choice, they just draw the line in different places as to the extent of subjectivity. Strict subjectivists allow most choice, drawing the line at infringements of the axioms of probability.[2] Proponents of empirically based subjective probability occupy a halfway house, allowing extensive choice but insisting that evidence of physical probabilities as well as the axioms of probability constrain degrees of belief. Objective Bayesians go furthest by also using equivocation considerations to narrow down the class of acceptable degrees of belief.

9.3.2 Fully objective probability

We see, then, that objectivity is a matter of degree and that while subjectivity may infect some problems, objective Bayesianism yields a relatively high degree of objectivity. We have been focusing on what might be called *epistemic objectivity*, the extent to which an agent's degrees of belief are determined by her evidence, both explicit ($\mathcal{E}$) and implicit (embodied in her choice of language $\mathcal{L}$). In applications of probability, a high degree of epistemic objectivity is an important desideratum: disagreements as to probabilities can be attributed to

---
[2]Subjectivists usually advocate a few further constraints: for example, known truths must be given probability 1 and degrees of belief should be updated by conditionalization.

differences in evidence or language; by agreeing on evidence and language, consensus can be reached on probabilities.

While epistemic objectivity seeks uniqueness relative to evidence and language, there are stronger grades of objectivity. In particular, the strongest grade of objectivity, *full objectivity*, that is, uniqueness simpliciter, arouses philosophical interest. Are probabilities uniquely determined, independently of evidence and language? If two agents disagree as to probabilities must at least one of them be wrong, even if they differ as to evidence and language? Intuitively, many probabilities are fully objective: there seems to be a fact of the matter as to the probability that an atom of cobalt-60 will decay in 5 years, and there seems to be a fact of the matter as to the chance that a particular roulette wheel will yield a black on the next spin. (A qualification is needed. Chances cannot be quite fully objective inasmuch as they depend on time. There might now be a probability just under 0.5 of cobalt-60 atom decaying in the next 5 years; after the event, if it has decayed its chance of decaying in that time frame is 1.

Arguably, objective Bayesianism has the wherewithal to account for intuitions about full objectivity. By considering some ultimate evidence $\hat{\mathcal{E}}$ and ultimate language $\hat{\mathcal{L}}$, one can define the fully objective probability $\hat{P} = P_{\hat{\mathcal{E}}}^{\hat{\mathcal{L}}}$ in terms of the degrees of belief one ought to adopt if one were to have this ultimate evidence and language. This is the *ultimate belief* notion of probability; a similar view was put forward by Johnson (1932c, pp. 10–11) in the context of a logical interpretation of probability.

What should be included in $\hat{\mathcal{E}}$? Clearly, it should include all information relevant at time t. To be on the safe side, we can take $\hat{\mathcal{E}}$ to include all fundamental facts about the universe that are determined by time t—the entire fundamental history of the universe (the pattern of instantiation of fundamental properties and relations by particular things) up to and including time t. (Remember: the quest for a fully objective notion of single-case probability is of philosophical rather than practical interest.) Correspondingly, the ultimate language $\hat{\mathcal{L}}$ needs to be able to express any proposition to which probabilities might attach. Thinking in terms of a predicate language, it should have predicate symbols corresponding to all the fundamental properties and relations, the means to define non-fundamental features, and an individual constant for each distinguishable object.

This yields a (close to) fully objective concept of probability, the ultimate belief interpretation. But one might wish to go further by claiming that chance itself should be analysed in terms of this notion of probability. This yields an ultimate belief interpretation of chance: the chances are just the rational degrees of belief of an agent with ultimate evidence and language.

While the ultimate belief interpretation of chance is relatively straightforward to state, much needs to be done to show that this type of approach is viable. One needs to show that this notion can capture our intuitions about chance. Moreover, one needs to show that the account is coherent—in particular, one might have concerns about circularity: if probabilistic beliefs are beliefs about probability, yet probability is defined in terms of probabilistic beliefs, then probability appears to be defined in terms of itself.

But this apparent circularity dissolves when we examine the premises of this circularity argument more closely. Indeed, at most one premise can be true. In our framework, 'probability is defined in terms of probabilistic beliefs' is true if we substitute 'chance' for 'probability' and 'rational degrees of belief' for 'probabilistic beliefs': chance is defined in terms of rational degrees of belief. But then the first premise is false. Rational degrees of belief are not beliefs about chance, they are *partial* beliefs. According to this reading 'probabilistic' modifies 'belief', isolating a type of belief; it does not specify the object of belief. On the other hand, if the first premise is to be true and 'probabilistic beliefs' are construed as beliefs about probability, then the second premise is false since chance is not here defined in terms of beliefs about probability. Thus, neither reading permits the conclusion that probability is defined in terms of itself.

Note that Bayesian statisticians often consider probability distributions over probability parameters. These *can* sometimes be interpreted as degrees of belief about chances, where chances are special degrees of belief. But there is no circularity here either. This is because the degrees of belief about chances are of a higher order than the chances themselves. Consider, for instance, a degree of belief that a particular coin toss will yield heads. The present chance of the coin toss yielding heads can be defined using such degrees of belief. One can then go on to formulate the higher-order degree of belief that the chance of heads is 0.5. But this degree of belief is not used in the (lower-order) definition of the chance itself, so there is no circularity. (One can go on to define progressively higher-order chances and degrees of belief—regress, rather than circularity, is the obvious problem.)

One can make a stronger case for circularity though. The Calibration norm says that degrees of belief ought to be set to chances where there is evidence of chances. Under such a reading, the concept of rational degree of belief appeals to the notion of chance, yet in this section chances are being construed as special degrees of belief; circularity again. Here circularity is not an artifice of ambiguity of terms like 'probabilistic beliefs'. However, as before, circularity does disappear under closer investigation. One way out is to claim that there are two notions of chance in play: a physical notion which is used in the Calibration norm and an ultimate belief notion which is defined in terms of degrees of belief. But this strategy would not appeal to those who find a physical notion of chance metaphysically or epistemologically dubious. An alternative strategy is to argue that the notion of chance in the formulation of the Calibration norm is simply *eliminable*. One can substitute references to chance with references to the *indicators* of chance instead. Intuitively, symmetry considerations, physical laws and observed frequencies all provide some evidence as to chances; one can simply say that an agent's degrees of belief should be appropriately constrained by her evidence of symmetries, laws, and frequencies. While this may lead to a rather more complicated formulation of the Calibration norm, it is truer to the epistemological route to degrees of belief—the agent has direct evidence of the indicators of chance rather than the chances themselves. Further, it shows how these indicators of chance can actually provide evidence for chances: evidence of frequencies constrains degrees of belief, and chances are just special degrees of belief. Finally, this strategy eliminates circularity, since it shows how degrees of belief can be defined independently of chances.

The ultimate belief notion of chance is not quite fully objective: it is indexed by time. Moreover, we saw above that various kinds of non-uniqueness can creep in, and as to which functions are sufficiently equivocal will be relative to the operating context. Nevertheless, this ultimate belief notion of chance is arguably more objective than the influential physical notion of chance put forward by David Lewis (1980, 1994). We saw in §3.3 that Lewis put forward a version of the Calibration norm which he called the *Principal Principle*: evidence of chances ought to constrain degrees of belief. However, Lewis did not go on to advocate the ultimate belief notion of chance presented here: 'chance is [not] the credence warranted by our total available evidence ... if our total evidence came from misleadingly unrepresentative samples, that wouldn't affect chance in any way' (Lewis 1994, p. 475). Unrepresentative samples are perhaps not such a problem for the ultimate belief approach, because the entire history of the universe up to the time in question is likely to contain more information pertinent to an event than simply a small sample frequency—plenty of large samples of relevant events, and plenty of relevant qualitative information, for instance. (However, when t is small $\hat{\mathcal{E}}$ will offer little to go on, being unrepresentative of the way things turn out in the future history of the universe. So very shortly after the big bang, the ultimate belief notion of chance yields more equivocal chances than will perhaps a physical notion of chance.)

Lewis instead took chances to be products of the best system of laws, the best way of systematizing the universe. The problem is that the criteria for comparing systems of laws—a balance between simplicity and strength—seem to be highly subjective. What counts as simple for a rocket scientist may be complicated for a robot and vice versa. (In response Lewis 1994, p. 479 just plays the optimism card: 'if nature is kind to us, the problem needn't arise'.) This is not a problem that besets the ultimate belief account: as Lewis accepts, there does seem to be a fact of the matter as to how evidence should inform degrees of belief. Thus, an ultimate belief notion of chance, despite being an epistemic rather than a physical notion, suffers from a less problematic kind of subjectivity than Lewis' theory.

Note that Lewis' approach also suffers from a type of circularity known as *undermining*. Because chances for Lewis are analysed in terms of laws, they depend not only on the past and present state of the universe, but also on the future of the universe: 'present chances are given by probabilistic laws, plus present conditions to which those laws are applicable, and ... those laws obtain in virtue of the fit of candidate systems to the whole of history' (Lewis 1994, p. 482).

Of course, non-actual futures (i.e. series of events which differ from the way in which the universe will actually turn out) must have positive chance now, for otherwise the notion of chance would be redundant. Thus, there is now a positive chance of events turning out in the future in such a way that present chances turn out differently. But this yields a paradox: present chances cannot turn out differently to what they actually are. Lewis (1994) has to modify the Principal Principle to avoid a formal contradiction, but this move does not resolve the intuitive paradox. In contrast, under the ultimate belief account present chances depend on just the past and the present state of the universe, not the future, so present chances cannot undermine themselves.

10

Objective Bayesianism in Perspective

In this chapter, we take stock and consider some questions and applications that merit future attention. In §10.1, we encounter some questions arising from the topics covered in this book. Sections 10.2, 10.3, and 10.4 advertise the application of objective Bayesian epistemology to statistics, confirmation theory, and metaphysics, respectively. As we see, there is plenty on the agenda for those wishing to contribute to the objective Bayesian research programme.

10.1 The state of play

Objective Bayesianism is not an easy theory to defend. While strictly subjective Bayesianism appeals to the Probability norm and empirically based subjectivism invokes a Calibration norm, objective Bayesianism also subscribes to an Equivocation norm, and this last norm bears the brunt of much of the criticism levelled at objective Bayesianism. But the basic intuition behind this norm is sound: if someone's evidence leaves the truth or falsity of θ open, then she would be irrational to strongly believe θ or its negation. This book is written in the belief that it is better to contribute to the struggle to state and defend the right position than to settle for a more easily defensible position that is only a part of the story. This struggle has barely begun, however; in particular, there are many ways in which the approach of this book can be developed further, as we now see.

Chapter 2 placed objective Bayesian epistemology within the context of the ongoing search for a viable interpretation of probability. Since the focus of the book is on objective Bayesian epistemology rather than interpretations of probability, we merely scratched the surface here. It was suggested that objective Bayesianism compares favourably with other interpretations with respect to the desiderata of §2.1: Objectivity, Calculi, Epistemology, Variety, and Parsimony. It remains to provide a sustained defence of this claim and to explore other desiderata. (This task is likely to require a book in itself!) The chapter also stated the core objective Bayesian formalism on propositional languages. Pragmatic questions—for example, when does a belief function count as *sufficiently* equivocal?—have not been taken up in this book and are open to further investigation.

Chapter 3 offered some pragmatic motivation for the norms of objective Bayesianism. We saw that each norm admits certain qualifications. The Probability norm should be circumscribed if there are angels guarding an agent from the consequences of her beliefs or if the agent needs to form beliefs about propositions which she cannot be expected to know are logically related (§3.2). The Calibration norm should be modified if evidence other than that of chances and of influence relationships can constrain degrees of belief (§3.3). The Equivocation norm should be modified if it turns out that the divergence measure explored in this book is not appropriate to the loss function for the agent's current operating context (§3.4). It would be interesting to explore these qualifications in more detail and to investigate any further qualifications that may arise.

Chapter 4 argued that the objective Bayesian approach to updating degrees of belief in the light of new evidence is preferable to conditionalization, the leading subjective Bayesian method. One consequence of this is that conditional probabilities have a less important status under objective Bayesianism than they do under subjective Bayesianism: for subjectivists $P(\theta|\varphi)$ represents the degree to which an agent should believe θ were she to learn φ, but this is not always so for objectivists (§4.4). This fairly radical departure from orthodoxy will have repercussions for objective Bayesian method—repercussions that would be well worth elucidating in more detail.

Chapter 5 extended the objective Bayesian formalism from propositional to predicate languages. Some of the properties of the extended system were highlighted, but this is new ground and there are many more questions to explore. For example, when is $\downarrow\mathbb{E}$ closed? If $\mathcal{E}$ takes the form of an arbitrary set of sentences of $\mathcal{L}$, what form does $\downarrow\mathbb{E}$ take? How should the formalism be extended to handle more general languages such as languages with equality or higher-order languages?

Chapter 6 introduced the objective Bayesian net (obnet) formalism and its generalization to objective credal nets. We saw that this computational machinery can be applied to calculate objective Bayesian probabilities in areas (such as cancer prognosis) that admit a wide variety of evidence. But so far implementational issues have not been addressed in any detail; they need to be if efficient and user-friendly software tools are to be developed.

Chapter 7 applied objective Bayesianism to the problem of providing a semantics for probabilistic logic. The objective Bayesian semantics is but one of a range of interpretations, but the credal net formalism can be exploited as a computational framework for all these interpretations. Again, implementation is a natural next step. As is relaxing of assumptions, in particular the regularity assumption to which we appealed.

Chapter 8 applied objective Bayesianism to the problem of aggregating the judgements of a number of agents. It put forward the following judgement aggregation procedure: first merge the evidence of the original agents; next apply objective Bayesianism to calculate the degrees of belief than an agent with this merged evidence should adopt; finally judge those propositions whose judgement maximizes expected utility. Section 8.5 mentioned some open issues: it remains to explore the properties of the suggested merging operator; it also remains to say in more detail how utilities are to be best aggregated in the context of this kind of solution.

Chapter 9 extended the objective Bayesian formalism to the language of the mathematical theory of probability, argued that some language relativity is to be expected, and discussed questions of objectivity, including an ultimate belief interpretation of probability. We saw that on the richer language of probability theory, implementing the Equivocation norm requires some thought. This is an active area of research amongst objective Bayesian statisticians, but one that is calling out for some unification and consolidation. Turning to language relativity, the question is now not whether objective Bayesianism displays language relativity, but whether it displays the *right amount* of language relativity. This question still needs answering; in the absence of an answer the onus is on the critic of objective Bayesianism to show that such relativity as there is is pernicious. The ultimate belief notion of chance is a promising line of research, but much more needs to be done to show it is really viable. For example, is it a substantial problem for the ultimate belief interpretation that it deems chances to be very different in the first moments of the universe?

This book has hopefully shown that objective Bayesianism is well motivated and a promising area of research. While the future may yield cogent criticisms of objective Bayesianism, current criticisms are surmountable. On current evidence, one should follow the norms of objective Bayesianism.

We now consider some further items on the agenda of the objective Bayesian research programme: understanding statistics, explicating scientific confirmation, and extrapolating the objective Bayesian approach to entities other than probability.

10.2 Statistics

The orthodox view in the philosophy of statistics goes something like this: there are three principal paradigms of statistics—the frequentist (also known as 'classical'), the subjectivist, and the objective Bayesian approach—at most one of which can be right. Philosophers of statistics tend to think of objective Bayesian statistical methods as subjective methods plus objective prior probabilities, and tend to view the task of determining objective prior probabilities as hopeless. Frequentist statistics is often viewed as a hotchpotch of disparate methods, lacking a coherent unifying framework. Hence, subjectivism is typically viewed by philosophers as being on the firmest ground.

On the other hand, many practising statisticians—reasonably enough—do not want the conclusions of their research to be classed as 'subjective'. They often take a pragmatic stance, shying away from foundations and calling on whichever methods are best suited to (or most developed for) the application of the moment.

If the framework of this book is taken seriously, the orthodox view in the philosophy of statistics should be overturned. The different paradigms are not mutually exclusive but are complementary in the sense that subjective statistical methods shed light on the content of the Probability norm, frequentist statistical methods determine which probability models (functions) are compatible with evidence and thus implement aspects of the Calibration norm, and objective Bayesian statistical methods explore issues to do with the Equivocation norm. In the objective Bayesian framework of this book, the norms fit together and so one might expect methods from the different paradigms to be used as a part of an objective Bayesian statistical analysis.

Objective Bayesianism is not just subjectivism plus objective priors. First and foremost, while objective Bayesian updating often agrees with subjective Bayesian updating—that is, conditionalization—there can be differences (see Chapter 4), so neither the use of conditionalization, nor therefore the use of Bayes' theorem, should be taken as a defining feature of Bayesian statistics. But there are other differences too. Not least, subjective choice, introspection, and elicitation play less of an epistemological role in determining probabilities on the objective Bayesian account.

The task of determining objective Bayesian probabilities is difficult, but not hopeless. It gets harder as we move from a finite propositional language to a predicate language to the language of the mathematical theory of probability, and there is room for some indeterminacy (and hence subjectivity) to creep in at each stage. But this is just subjectivity, not inconsistency or paradox as is often charged.

That frequentist methods sometimes seem disparate does not make them bad methods. In fact, they play a crucial role in the objective Bayesian scheme. If evidence consists of data from a large number of trials that are identically and independently distributed (iid), then since determining which probability functions are compatible with that evidence is a precondition for determining the sufficiently equivocal probability functions that are compatible with that evidence, frequentist methods are unavoidable. But, just as the current operating context has a bearing on which probability functions get classed as *sufficiently* equivocal, it can also have a bearing on which probability functions are ruled out as incompatible. In particular, frequentist significance levels may depend on contextual factors such as the risks attached to erroneously ruling out the correct probability function. That decisions need to be made about what counts as sufficiently equivocal and which significance level should be used means that there is further room for subjectivity to creep in. There is no suggestion that in every particular case there will be a single optimal significance level or a single optimal cut-off point for sufficiently equivocal functions. But these decisions are open to reasoned debate and non-uniqueness is no grounds for excluding them from rational epistemology.

The framework of this book, then, does not preclude the practising statistician from employing methods from different statistical paradigms. In fact, the practising statistician *ought* to bridge the paradigms, applying frequentist methods to the task of determining which probability functions are compatible with evidence and applying objective Bayesian methods to the task of determining which of those functions should be used as a basis for decision and action.

10.2.1 *Evidential probability and objective Bayesianism*

As an example of the way in which frequentist statistics and objective Bayesian epistemology might mesh, we take a quick look at a possible alliance between evidential probability and objective Bayesianism.

The Calibration norm has been explicated as $P_\mathcal{E} \in \mathbb{E} = \langle \mathbb{P}^* \rangle \cap \mathbb{S}$, where $\langle \mathbb{P}^* \rangle$ is the convex hull of the set in which quantitative evidence suggests that the chance function P^* might lie and where $\mathbb{S}$ is a set of functions satisfying structural constraints imposed by qualitative evidence. The determination of $\langle \mathbb{P}^* \rangle$ is left as a question for statistics to answer. Now evidential probability, introduced in §7.2, is a statistical theory that aims to answer just such a question: what implications does evidence of frequencies have for other probabilities? As we saw, it contains rules for handling frequency intervals and reference classes, associating an interval with any sentence θ.

We also saw that such an interval can be construed, via the theory of second-order evidential probability, as bounds on the point-valued probability $P(\theta)$. These bounds are always convex, hence second-order evidential probability can be thought of as identifying the set $\langle \mathbb{P}^* \rangle$ with the set of probability functions that satisfy the evidential-probability consequences of the evidence. (Note that some consistency-maintenance procedure will need to be invoked to handle those cases in which evidential-probability consequences of evidence are unsatisfiable; one might take $\langle \mathbb{P}^* \rangle$ to include any probability function that satisfies the consequences of some maximal satisfiable subset of the evidence, for instance.)

One can then go on to apply objective Bayesianism as usual to determine a sufficiently equivocal probability function, from those that are compatible with the evidence. This embedding of evidential probability into the objective Bayesian method yields what might be called *EP-calibrated objective Bayesianism*, where EP stands for evidential probability (Wheeler and Williamson 2010, §5).

It might appear that there is a gap in the above scheme: evidential probability takes statistical statements of the form $\%(\theta(x), \rho(x), [l, u])$ as its starting point, but where do these statistical statements come from? Clearly, an important task of statistics is the inference of these statistical statements from data, and something needs to be said about this key step. In fact, the theory of evidential probability does say something about this step (Kyburg Jr and Teng 2001, §§11.3–11.5), which may be summarized as follows. Frequentist statistics warrants inferences of the form: given data and assumptions about the data-collection process and the underlying populations, the probability that $\%(\theta(x), \rho(x), [l, u])$ is in some set X. Evidential probability provides a threshold δ for acceptance, and one should accept the statistical statement $\%(\theta(x), \rho(x), [l, u])$, from all those whose probability surpasses the threshold—that is, from all those for which $X \subseteq [1 - \delta, 1]$—, whose interval $[l, u]$ is the narrowest. In our framework, this may be construed as an assertion about evidence—about what you should grant: if you grant the data and the relevant assumptions, then you should grant that $\%(\theta(x), \rho(x), [l, u])$. So frequentist statistics is required to help determine which statistical statements should be taken as evidence, as well as to help determine the set of probability functions compatible with that evidence. Of course, frequentist statistics itself does not normally provide a threshold of acceptance—this is something that the agent brings to the inferential problem and is constrained by considerations to do with the agent's current operating context (e.g. the risks attached to erroneous acceptance or rejection). It may be that in a particular context there is no suitable threshold of acceptance for a statistical statement so the question of its unqualified acceptance does not arise. In which case it is the original claim of the form $\%(\theta(x), \rho(x), [l, u])^X$, ascribing the set X of probabilities to the statistical statement, that should be taken as evidence; such claims can be handled directly by second-order evidential probability (§7.2).

We saw in §7.2 that credal nets provide a calculus for second-order evidential probability, and in Chapter 6 and §7.4 that they also provide a calculus for objective Bayesian probability. In fact, the two kinds of net can be combined to yield a calculus for EP-calibrated objective Bayesianism.

Consider the following rather trivial example, expressed as a question in probabilistic logic:

$$\%x(\theta(x), \rho(x), [.2, .5]), \rho(t), \forall x(\theta(x) \rightarrow \psi(x))^{3/4} \approx \psi(t)^{?}$$

Now the first two premises yield $\theta(t)^{[.2,.5]}$ by applying the rules of evidential probability. This constraint combines with the third premise to yield an answer to the above question by appealing to objective Bayesianism. This answer can be calculated by constructing the following credal net:

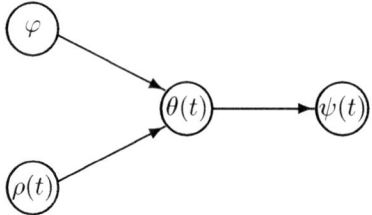

Here φ is the first premise. The left-hand side of this net is the second-order evidential probability net, with associated probability constraints

$$P(\varphi) = 1, P(\rho(t)) = 1,$$
$$P(\theta(t)|\varphi \wedge \rho(t)) \in [.2, .5], P(\theta(t)|\neg\varphi \wedge \rho(t)) = 0$$
$$P(\theta(t)|\varphi \wedge \neg\rho(t)) = P(\theta(t)|\neg\varphi \wedge \neg\rho(t)) = 0.$$

The right-hand side of this net is the obnet with associated probabilities:

$$P(\psi(t)|\theta(t)) = 7/10, P(\psi(t)|\neg\theta(t)) = 1/2.$$

Standard inference algorithms then yield an answer of $7/12$ to our question:

$$\%x(\theta(x), \rho(x), [.2, .5]), \rho(t), \forall x(\theta(x) \rightarrow \psi(x))^{3/4} \approx \psi(t)^{7/12}.$$

Hence, there is no barrier in principle to employing frequentist statistical methods to implement the Calibration norm, and since the credal net computational method is so general, there is still scope for using it to calculate objective Bayesian probabilities. Here evidential probability can be used to isolate the probability functions compatible with evidence, at which point objective Bayesian methods are applied to determine which probability function should be used as a basis for action. Interestingly, this scheme was not so far from Kyburg's own view (Kyburg Jr 1970). While Kyburg rejected the betting interpretation of belief, he did suggest that evidential probabilities might, via a version of the Principal Principle, constrain (but not fully determine) the choice of a probability function to be adopted as a basis for action. Kyburg viewed statistics as the combined enterprise incorporating both the assessment of evidential impact and the determination of probability as a basis for action, with evidential probability fulfilling the former role and Bayesian statistics pertaining to the latter task.

10.3 Confirmation and science

One of the success stories of Bayesian epistemology has been its application to confirmation theory, in particular to providing an account of the confirmation of scientific hypotheses (see, e.g. Carnap 1950; Howson and Urbach 1989a). The core idea is to explicate the degree to which evidence $\mathcal{E}$ confirms a hypothesis θ to be the degree to which an agent with evidence $\mathcal{E}$ ought to believe θ, with this degree of belief understood in terms of Bayesian probability.

Under a subjectivist construal, total evidence $\mathcal{E}$ confirms θ to degree $P(\theta|\varepsilon)$ where P is the chosen prior belief function of some particular agent with evidence ε, as long as $\mathcal{E}$ can be captured by a sentence ε of the agent's language. Under an objective Bayesian construal, total evidence $\mathcal{E}$ confirms θ to degree $P_{\mathcal{E}}(\theta)$ where $P_{\mathcal{E}} \in \Downarrow\mathbb{E}$ is the belief function of an agent with evidence $\mathcal{E}$. In either case, the degree of confirmation is relative to agent, but in the objective Bayesian case there is typically far less room for subjective choice of belief function.

There are two main roles for the application of Bayesian confirmation theory to science. The first is retrospective: to reconstruct and rationalize past progress in science. The second is prospective: to advise scientists and to devise machines that can automate aspects of the scientific analysis and discovery process.

The first, retrospective, role is fraught with difficulty. It is hard enough to get a grasp on what our current total evidence is, let alone to try to reconstruct the total evidence available to scientists at a particular point far in the past. It is also very hard to reconstruct the language and conceptualization of past scientists: while this is the bread and butter of historians of science, few would claim to have achieved such success as to permit a full-blown quantitative Bayesian analysis. Finally, with the retrospective role there is no real need for the precision offered by Bayesian accounts. What are of interest are the qualitative lessons to be learnt—for example, were scientists right to dismiss such-and-such a theory?—and applying a quantitative confirmation theory is tantamount to using sledgehammer to crack a nut.

The second, prospective, role is more fruitful. The evidence currently available to any particular research group is well circumscribed and largely made explicit, even if it is not all recorded for posterity or for future retrospective applications of Bayesian confirmation theory. A group's current conceptualization of their domain is also in principle available to them, and capable of being made explicit with a bit of work. Finally, a quantitative analysis can be very helpful: it can be useful to know the precise extent to which a range of possible experiments might confirm a given theory; if one is programming a machine to analyse experiments the last thing one needs is a vaguely formulated qualitative lesson—numbers must be crunched.

Objective Bayesianism has several things going for it as a foundation for this prospective role. First, its norms are well suited to confirmation as well as belief. The axioms of probability are plausible constraints on confirmation: for example, it is natural to require that degrees of confirmation attaching to a hypothesis and its negation sum to 1. Calibration makes explicit the requirement that evidence should constrain confirmation. And if evidence leaves open the probability of a hypothesis, one does not want to say that evidence confirms the hypothesis to a high (or a low) degree—confirmation is more equivocal than that. Second, objective Bayesian updating does not suffer from the limitations of conditionalization outlined in §4.3. For example, there is no need to formulate every possible future sequence of evidence in one's current language, and there is no need to specify probabilities conditional on each such sequence. Indeed, objective Bayesianism can cope well with language change (see, e.g. Williamson 2005a, chapter 12). Third, objective Bayesianism is (more rather than less) objective. While belief is a concept that smacks of subjectivity, confirmation has more objective connotations and an objective Bayesian approach can account for this.

10.4 Epistemic metaphysics

This book sets out with an epistemological question: to what extent should an agent believe the various propositions expressible in her language? We have seen that an answer can be constructed that relies heavily on the formal notion of probability: rational degrees of belief are formally probabilities. Thus, objective Bayesianism provides one interpretation of single-case probability. It is most naturally employed as a notion of probability that is relativized to evidence and language, but it also serves as an interpretation of unrelativized probability according to which probabilities are the degrees of belief that an agent with ultimate evidence and language ought to adopt (§9.3). So we have seen that a theory constructed in answer to an epistemological question about rational belief can also provide an answer to a metaphysical question about the nature of probability. Since rational degrees of belief are probabilities, at least some probabilities are rational degrees of belief.

This duality between epistemology and metaphysics extends to scenarios other than that of rational degree of belief and probability: an epistemological theory about which beliefs qualify as rational can often be translated into a metaphysical analysis. An analysis of a particular kind of entity X in terms of rational X-beliefs can be called an *epistemic metaphysics* of X. In this book,

X is probability and the rational probabilistic beliefs in question are rational degrees of belief. But epistemic theories can be developed for other problematic kinds of entity too. An epistemic metaphysics of causality, for instance, analyses causal relationships in terms of rational causal beliefs (Williamson 2005a, chapter 9). Here causal beliefs—just like degrees of belief—allow an agent to draw inferences and construct explanations. But, unlike the probabilistic case, a causal belief is a directed, qualitative sort of belief rather than a quantitative belief. While a degree of belief is naturally represented by an expression of the form $P(\theta) = x$, where x is a real number in the unit interval, a causal belief may be represented by an expression of the form $A_i \longrightarrow A_j$ signifying that A_i is a cause of A_j. Similarly, while the entirety of an agent's rational degrees of belief may be captured by a probability function, the entirety of her rational causal beliefs may be captured by a directed acyclic graph (dag). Depending on the agent's evidence and language, some set of dags will qualify as rational—that is, appropriate beliefs for an agent with that evidence and language—and the other graphs will be deemed irrational. It is an important epistemological question as to which graphs are rational and which are irrational, and an epistemological theory of causality might be expected to segregate the rational from the irrational. Now, any such epistemological theory generates a dual metaphysical theory of causality, the corresponding *epistemic metaphysics* of causality. According to such a theory, the causal relation is to be analysed as the set of causal beliefs that an agent with ultimate evidence and language ought to adopt. Causality, then, like probability, can be construed as epistemic yet largely objective: under an epistemic metaphysics, causal relations, like probabilistic relations, are not 'out there' to be perceived, they are instead a feature of the way we structure our inferences; but rational considerations strongly constrain appropriate causal (and probabilistic) relations so they are not simply a question of subjective choice.

A similar view can be had about mathematical entities: the entities of mathematics are arguably analysable in terms of the mathematical beliefs of an agent with ultimate evidence and language, and under such a view they are epistemic yet largely objective (Williamson 2009c, part IV). While degrees of belief quantify strength of conviction and causal beliefs systematize the inferences we draw between occurrences of contingent events, mathematical beliefs have yet a different role, systematizing the inferences that follow necessary connections concerning basic concepts such as number, set, function, category, and space. Given the current language of mathematics and given what we currently grant (definitions, axioms, rules of inference, conjectures, and so on), certain sets of mathematical beliefs are appropriate while others are inappropriate; it is up to an epistemological theory of mathematics to separate the sheep from the goats. Such an epistemological theory yields a metaphysical dual: the view that the mathematical truths are just the rational mathematical beliefs of an agent with ultimate evidence and language, and their truth-makers are the epistemological facts about how evidence and language consort to constrain rational mathematical beliefs. If one holds such a view there is no need to posit a world of abstract mathematical entities, totally separate from our own world yet somehow knowable by perception or inference. But disavowing this realm of entities need not be accompanied by a loss of objectivity since standards of rationality (standards of proof and precision together with other desiderata such as applicability inside or outside mathematics) strongly constrain what is, and what is not, appropriate as a mathematical belief.

So the theory of objective Bayesianism can act as a blueprint for other epistemological theories and each epistemological theory will have its metaphysical dual, an epistemic metaphysics. If some X is problematic in the sense that it does not appear to be straightforwardly analysable in terms of worldly features, this may be because X is in fact epistemic. In which case an epistemic metaphysics of X is required, and that theory will need to invoke norms that constrain rational X-beliefs. In sum, the lessons of objective Bayesianism can be more widely applied, and the programme of developing a viable theory of objective Bayesianism fits into a more general programme of providing epistemic metaphysics for epistemic entities.

References

Abraham, J. A. (1979). An improved algorithm for network reliability. *IEEE Transactions on Reliability*, **28**, 58–61.

Aczél, J. and Daróczy, Z. (1975). *On measures of information and their characterizations*. Academic Press, New York.

Alchourrón, C. E., Gärdenfors, P., and Makinson, D. (1985). On the logic of theory change: partial meet functions for contraction and revision. *Journal of Symbolic Logic*, **50**, 510–30.

Amari, Shun-ichi and Nagaoka, Hiroshi (1993). *Methods of information geometry* (2000 edn). American Mathematical Society/Oxford University Press, Providence, RI.

Bacchus, Fahiem, Kyburg, Jr., Henry E., and Thalos, Mariam (1990). Against conditionalization. *Synthese*, **84**(3), 475–506.

Baillie, Patricia (1973). Confirmation and the Dutch book argument. *The British Journal for the Philosophy of Science*, **24**(4), 393–7.

Barnett, O. W. and Paris, J. B. (2008). Maximum entropy inference with quantified knowledge. *Logic Journal of the IGPL*, **16**, 85–98.

Bayes, Thomas (1764). An essay towards solving a problem in the doctrine of chances. *Philosophical Transactions of the Royal Society of London*, **53**, 370–418.

Bernoulli, Jakob (1713). *Ars Conjectandi* (2006 edn). The Johns Hopkins University Press, Baltimore. Translated by Edith Dudley Sylla.

Billingsley, Patrick (1979). *Probability and measure* (Third (1995) edn). Wiley, New York.

Borel, Emile (1943). *Probabilities and life* (1962 edn). Dover, New York.

Bovens, Luc and Rabinowicz, Wlodek (2006). Democratic answers to complex questions-an epistemic perspective. *Synthese*, **150**, 131–53.

Carnap, Rudolf (1950). *Logical foundations of probability*. Routledge and Kegan Paul, London.

——(1952). *The continuum of inductive methods*. University of Chicago Press, Chicago, IL.

Chopra, Samir, Ghose, Aditya, and Meyer, Thomas (2006). Social choice theory, belief merging, and strategy-proofness. *Information Fusion*, **7**(1), 61–79.

Corfield, David (2001). Bayesianism in mathematics. In *Foundations of Bayesianism* (eds. D. Corfield and J. Williamson), pp. 175–201. Kluwer, Dordrecht.

Cover, Thomas M. and Thomas, Joy A. (1991). *Elements of information theory*. Wiley, New York.

Cox, R. T. (1946). Probability, frequency and reasonable expectation. *American Journal of Physics*, **14**(1), 1–3.

Dawid, A. P. (1982). The well-calibrated Bayesian. *Journal of the American Statistical Association*, **77**, 604–13. With discussion.

de Cooman, Gert and Miranda, Enrique (2007). Symmetry of models versus models of symmetry. In *Probability and Inference: Essays in Honour of Henry E. Kyburg Jr.* (eds. W. Harper and G. Wheeler), pp. 67–149. College Publications, London.

de Finetti, Bruno (1937). Foresight: its logical laws, its subjective sources. In *Studies in Subjective Probability* (Second (1980) edn) (eds. H. E. Kyburg and H. E. Smokler), pp. 53–118. Robert E. Krieger Publishing Company, Huntington, New York.

Deakin, Michael A. B. (2006). The wine/water paradox: background, provenance and proposed resolutions. *The Australian Mathematical Society Gazette*, **33**, 200–5.

del Val, Alvaro and Shoham, Yoav (1994). A unified view of belief revision and update. *Journal of Logic and Computation*, **4**(5), 797–810.

Delgrande, James P. and Schaub, Torsten (2007). A consistency-based framework for merging knowledge bases. *Journal of Applied Logic*, **5**(3), 459–77.

Dias, Penha Maria Cardoso and Shimony, Abner (1981). A critique of Jaynes' maximum entropy principle. *Advances in Applied Mathematics*, **2**(2), 172–211.

Dietrich, Franz and List, Christian (2007). Arrow's theorem in judgment aggregation. *Social Choice and Welfare*, **29**(1), 19–33.

Earman, John (1992). *Bayes or bust?* MIT Press, Cambridge, MA.

Fantl, Jeremy and McGrath, Matthew (2007). On pragmatic encroachment in epistemology. *Philosophy and Phenomenological Research*, **75**, 558–789.

Festa, Roberto (1993). *Optimum inductive methods: a study in inductive probability, Bayesian statistics, and verisimilitude*. Kluwer, Dordrecht.

Fienberg, Stephen E. (2006). When did Bayesian inference become "Bayesian"? *Bayesian Analysis*, **1**(1), 1–40.

Foley, Richard (1993). *Working without a net*. Oxford University Press, New York.

Franklin, James (2001). Resurrecting logical probability. *Erkenntnis*, **55**, 277–305.

Fridlyand, J., Snijders, A. M., Ylstra, B., Li, H., Olshen, A., Segraves, R., Dairkee, S., Tokuyasu, T., Ljung, B. M., Jain, A. N., McLennan, J., Ziegler, J., Chin, K., Devries, S., Feiler, H., Gray, J. W., Waldman, F., Pinkel, D., and Albertson, D. G. (2006). Breast tumor copy number aberration phenotypes and genomic instability. *BMC Cancer*, **6**, 96.

Friedman, Kenneth and Shimony, Abner (1971). Jaynes's maximum entropy prescription and probability theory. *Journal of Statistical Physics*, **3**(4), 381–4.

Gaifman, H. and Snir, M. (1982). Probabilities over rich languages. *Journal of Symbolic Logic*, **47**(3), 495–548.

Gauwin, Olivier, Konieczny, Sébastien, and Marquis, Pierre (2005). Conciliation and consensus in iterated belief merging. In *Proceedings of the 8th European Conference on Symbolic and Quantitative Approaches to Reasoning with Uncertainty*, pp. 514–26. Springer, Berlin.

Gillies, Donald (1991). Intersubjective probability and confirmation theory. *British Journal for the Philosophy of Science*, **42**, 513–33.

—— (2000a). *Philosophical theories of probability*. Routledge, London and New York.

—— (2000b). Varieties of propensity. *British Journal for the Philosophy of Science*, **51**, 807–35.

Good, I. J. (1952). Rational decisions. *Journal of the Royal Statistical Society. Series B*, **14**(1), 107–14.

Gottfried, Kurt and Wilson, Kennth G. (1997). Science as a cultural construct. *Nature*, **386**, 545–7. With discussion in vol. 387 p. 543.

Grove, Adam J. and Halpern, Joseph Y. (1997). Probability update: conditioning vs. cross-entropy. In *Proceedings of the 13th Annual Conference on Uncertainty in Artificial Intelligence*, pp. 208–14. Morgan Kaufmann, San Francisco, CA.

Grünwald, Peter (2000). Maximum entropy and the glasses you are looking through. In *Proceedings of the 16th Conference on Uncertainty in Artificial Intelligence*, pp. 238–46. Morgan Kaufmann. San Francisco, CA.

—— and Dawid, A. Philip (2004). Game theory, maximum entropy, minimum discrepancy, and robust Bayesian decision theory. *Annals of Statistics*, **32**(4), 1367–433.

Hacking, Ian (1971). Jacques Bernoulli's art of conjecturing. *British Journal for the Philosophy of Science*, **22**(3), 209–29.

Haenni, R. (2007). Climbing the hills of compiled credal networks. In *ISIPTA '07. 5th International Symposium on Imprecise Probabilities and Their Applications* (eds. G. de Cooman, J. Vejnarová, and M. Zaffalon), Prague, Czech Republic, pp. 213–22.

—— (2009). Probabilistic argumentation. *Journal of Applied Logic*, **7**(2), 155–76. Berlin.

Haenni, R., Kohlas, J., and Lehmann, N. (2000). Probabilistic argumentation systems. In *Handbook of Defeasible Reasoning and Uncertainty Management Systems* (eds. D. M. Gabbay and P. Smets), Volume 5: Algorithms for Uncertainty and Defeasible Reasoning, pp. 221–88. Kluwer Academic Publishers, Dordrecht.

—— Romeijn, Jan-Willem, Wheeler, Gregory, and Williamson, Jon (2010). *Probabilistic logic and probabilistic networks*. Synthese Library. Springer.

Halpern, Joseph Y. (1999a). A counterexample to theorems of Cox and Fine. *Journal of Artificial Intelligence Research*, **10**, 67–85.

Halpern, Joseph Y. (1999b). Cox's theorem revisited. *Journal of Artificial Intelligence Research*, **11**, 429–35.

——(2003). *Reasoning about uncertainty*. MIT Press, Cambridge, MA.

Halpern, Joseph Y. and Koller, Daphne (1995). Representation dependence in probabilistic inference. In *Proceedings of the 14th International Joint Conference on Artificial Intelligence (IJCAI 95)* (ed. C. S. Mellish), pp. 1853–60. Morgan Kaufmann, San Francisco, CA.

Harman, Gilbert (2004). Practical aspects of theoretical reasoning. In *The Oxford Handbook of Rationality* (eds. A. R. Mele and P. Rawling), pp. 45–56. Oxford University Press, Oxford.

——(2007). Epistemology as methodology. In *A Companion to Epistemology* (second edn) (eds. J. Dancy and E. Sosa). Blackwell. Oxford.

Harremoës, Poul, Gee, David, MacGarvin, Malcolm, Stirling, Andy, Keys, Jane, Wynne, Brian, and Vaz, Sofia Guedes (eds.) (2002). *The precautionary principle in the 20th century: late lessons from early warnings*. Earthscan, London.

Hartmann, Stephan, Pigozzi, Gabriella, and Sprenger, Jan (2007). Reliable methods of judgment aggregation. Technical Report 3593 (http://philsci-archive.pitt.edu/archive/00003593/), Philsci Archive.

Hawthorne, James and Makinson, David C. (2007). The quantitative/qualitative watershed for rules of uncertain inference. *Studia Logica*, **86**, 247–97.

Hilbert, David (1925). On the infinite. In *Philosophy of Mathematics: Selected Readings* (second edn) (eds. P. Benacerraf and H. Putnam). Cambridge University Press (1983), Cambridge.

Hild, Matthias, Jeffrey, Richard, and Risse, Mathias (2008). Preference aggregation after harsanyi. In *Justice, Political Liberalism, and Utilitarianism: Themes from Harsanyi and Rawls* (eds. M. Fleurbaey, J. A. Weymark, and M. Salles). Cambridge University Press, Cambridge.

Hobson, Arthur (1971). *Concepts in statistical mechanics*. Gordon and Breach, New York.

Hosni, H. and Paris, J. B. (2005). Rationality as conformity. *Synthese*, **144**(3), 249–85.

Howson, Colin (1997). Bayesian rules of updating. *Erkenntnis*, **45**, 195–208.

——(2000). *Hume's problem: induction and the justification of belief*. Clarendon Press, Oxford.

——(2001). The logic of Bayesian probability. In *Foundations of Bayesianism* (eds. D. Corfield and J. Williamson), pp. 137–59. Kluwer, Dordrecht.

——(2008). De finetti, countable additivity, consistency and coherence. *The British Journal for the Philosophy of Science*, **59**, 1–23.

——(2009). Can logic be combined with probability? Probably. *Journal of Applied Logic*, **7**(2), 177–87.

——— and Urbach, Peter (1989a). *Scientific reasoning: the Bayesian approach* (Second (1993) edn). Open Court, Chicago, IL.

——––(1989b). *Scientific reasoning: the Bayesian approach* (Third (2006) edn). Open Court, Chicago, IL.

Jaynes, E. T. (1957a). Information theory and statistical mechanics. *The Physical Review*, **106(4)**, 620–30.

——(1957b). Information theory and statistical mechanics II. *The Physical Review*, **108**(2), 171–90.

——(1973). The well-posed problem. *Foundations of Physics*, **3**, 477–92.

——(2003). *Probability theory: the logic of science*. Cambridge University Press, Cambridge.

Jeffreys, Harold (1931). *Scientific inference* (Second (1957) edn). Cambridge University Press, Cambridge.

——(1939). *Theory of Probability* (Third (1961) edn). Clarendon Press, Oxford.

Johnson, W. E. (1932a). Probability: axioms. *Mind*, **41**(163), 281–96.

——(1932b). Probability: the deductive and inductive problems. *Mind*, **41**(164), 409–23.

——(1932c). The relations of proposal to supposal. *Mind*, **41**(161), 1–16.

Kass, Robert E. and Wasserman, Larry (1996). The selection of prior distributions by formal rules. *Journal of the American Statistical Association*, **91**, 1343–70.

Katsuno, Hirofumi and Mendelzon, Alberto O. (1991). On the difference between updating a knowledge base and revising it. In *Proceedings of the Second International Conference on Principles of Knowledge Representation and Reasoning*, pp. 387–94. Morgan Kaufmann, San Francisco, CA.

Kelly, J. L. (1956). A new interpretation of information rate. *Bell Systems Technical Journal*, **35**, 917–26.

Keynes, John Maynard (1921). *A treatise on probability*. Macmillan (1948), London.

Kolmogorov, A. N. (1933). *The foundations of the theory of probability*. Chelsea Publishing Company (1950), New York.

Konieczny, Sébastien and Pino Pérez, Ramón (1998). On the logic of merging. In *Proceedings of the 6th International Conference on Principles of Knowledge Representation and Reasoning*, pp. 488–98. Morgan Kaufmann, San Francisco, CA.

——––(2002). Merging information under constraints: a logical framework. *Journal of Logic and Computation*, **12**(5), 773–808.

Kornhauser, L. A. and Sager, L. G. (1986). Unpacking the court. *Yale Law Journal*, **96**, 82–117.

Kraus, Sarit, Lehmann, Daniel, and Magidor, Menachem (1990). Nonmonotonic reasoning, preferential models and cumulative logics. *Artificial Intelligence*, **44**, 167–207.

Kullback, S. and Leibler, R. A. (1951). On information and sufficiency. *The Annals of Mathematical Statistics*, **22**, 79–86.

Kyburg Jr, Henry E. (1968). Bets and beliefs. *American Philosophical Quarterly*, **5**(1), 54–63.

—— (1970). Two world views. *Noûs*, **4**(4), 337–48.

—— (2003). Are there degrees of belief? *Journal of Applied Logic*, **1**, 139–49.

—— and Pittarelli, Michael (1996). Set-based bayesianism. *IEEE Transactions on Systems, Man, and Cybernetics*, **26**(3), 324–39.

Kyburg Jr, Henry E. and Teng, Choh Man (1999). Choosing among interpretations of probability. In *Proceedings of the 15th Annual Conference on Uncertainty in Artificial Intelligence (UAI-99)*, pp. 359–65. Morgan Kaufmann Publishers, San Francisco, CA.

—————— (2001). *Uncertain inference*. Cambridge University Press, Cambridge.

—————— and Wheeler, Gregory (2007). Conditionals and consequences. *Journal of Applied Logic*, **5**(4), 638–50.

Lange, Mark (1999). Calibration and the epistemological role of Bayesian conditionalization. *The Journal of Philosophy*, **96**, 294–324.

Laplace (1814). *A philosophical essay on probabilities*. Dover (1951), New York. Pierre Simon, marquis de Laplace.

Levi, Isaac (1980). *The enterprise of knowledge: an essay on knowledge, credal probability, and chance*. MIT Press, Cambridge, MA.

Levine, Raphael D. and Tribus, Myron (ed.) (1979). *The maximum entropy formalism*. MIT Press, Cambridge, MA.

Lewis, David K. (1980). A subjectivist's guide to objective chance. In *Philosophical Papers*, Volume 2, pp. 83–132. Oxford University Press (1986), Oxford.

—— (1994). Humean supervenience debugged. *Mind*, **412**, 471–490.

—— and —— (1998). Merging databases under constraints. *International Journal of Cooperative Information Systems*, **7**(1), 55–76.

List, C. and Pettit, P. (2002). Aggregating sets of judgments. An impossibility result. *Economics and Philosophy*, **18**, 89–110.

—— —— (2004). Aggregating sets of judgments. Two impossibility results compared. *Synthese*, **140**, 207–35.

Makinson, David (2005). *Bridges from classical to nonmonotonic logic*. King's College Publications, London.

Meyer, Thomas, Ghose, Aditya, and Chopra, Samir (2001). Social choice, merging and elections. In *Proceedings of the 6th European Conference on Symbolic and Quantitative Approaches to Reasoning with Uncertainty* (eds. Benferhat and Besnard), pp. 466–77. Springer. Volume 2143 of Lecture Notes in Artificial Intelligence. Berlin.

Mikkelson, Jeffrey M. (2004). Dissolving the wine/water paradox. *British Journal for the Philosophy of Science*, **55**, 137–45.

Miller, David (1966). A paradox of information. *British Journal for the Philosophy of Science*, **17**, 59–61.

Mongin, Phillippe (2001). The paradox of the Bayesian experts. In *Foundations of Bayesianism* (eds. D. Corfield and J. Williamson), pp. 309–38. Kluwer, Dordrecht.

Nagl, Sylvia, Williams, Matt, and Williamson, Jon (2008). Objective Bayesian nets for systems modelling and prognosis in breast cancer. In *Innovations in Bayesian networks: theory and applications* (eds. D. Holmes and L. Jain), pp. 131–167. Springer, Berlin.

Nau, Robert F. (2001). De Finetti was right: probability does not exist. *Theory and Decision*, **51**, 89–124.

Neapolitan, Richard E. (1990). *Probabilistic reasoning in expert systems: theory and algorithms*. Wiley, New York.

Nilsson, Nils J. (1986). Probabilistic logic. *Artificial Intelligence*, **28**, 71–87.

Oaksford, Mike and Chater, Nick (2007). *Bayesian rationality: the probabilistic approach to human reasoning*. Oxford University Press, Oxford.

Paris, J. B. (1994). *The uncertain reasoner's companion*. Cambridge University Press, Cambridge.

—— (2003). *Lecture notes on nonmonotonic logic*. Department of Mathematics, University of Manchester. www.maths.man.ac.uk/~jeff/lecture-notes/mt4181.ps.

—— and Vencovská, A. (1990). A note on the inevitability of maximum entropy. *International Journal of Approximate Reasoning*, **4**, 181–223.

—— —— (1997). In defence of the maximum entropy inference process. *International Journal of Approximate Reasoning*, **17**, 77–103.

—— —— (2001). Common sense and stochastic independence. In *Foundations of Bayesianism* (eds. D. Corfield and J. Williamson), pp. 203–40. Kluwer, Dordrecht.

—— —— (2003). The emergence of reasons conjecture. *Journal of Applied Logic*, **1**(3–4), 167–95.

Pauly, Marc and van Hees, Martin (2006). Logical constraints on judgement aggregation. *Journal of Philosophical Logic*, **35**, 568–85.

Pearl, Judea (1988). *Probabilistic reasoning in intelligent systems: networks of plausible inference*. Morgan Kaufmann, San Mateo, CA.

Pigozzi, Gabriella (2006). Belief merging and the discursive dilemma: an argument-based account to paradoxes of judgment aggregation. *Synthese*, **152**(2), 285–98.

Press, S. James and Tanur, Judith M. (2001). *The subjectivity of scientists and the Bayesian approach*. Wiley, Chichester.

Ramsey, Frank Plumpton (1926). Truth and probability. In *Studies in Subjective Probability* (Second (1980) edn) (eds. H. E. Kyburg and H. E. Smokler), pp. 23–52. Robert E. Krieger Publishing Company, Huntington, New York.

Ramsey, Frank Plumpton (1928). Reasonable degree of belief. In *Philosophical Papers* (1990 edn) (ed. D. H. Mellor). Cambridge University Press, Cambridge.

Reichenbach, Hans (1935). *The theory of probability: an inquiry into the logical and mathematical foundations of the calculus of probability* (1949 edn). University of California Press, Berkeley and Los Angeles. Translated by Ernest H. Hutten and Maria Reichenbach.

Rissanen, Jorma (1987). Stochastic complexity. *Journal of the Royal Statistical Society. Series B*, **49**(3), 223–39.

Rosenkrantz, Roger D. (1977). *Inference, method and decision: towards a Bayesian philosophy of science*. Reidel, Dordrecht.

Rowbottom, Darrell P. (2007). The insufficiency of the Dutch Book argument. *Studia Logica*, **87**, 65–71.

——(2008). On the proximity of the logical and 'objective Bayesian' interpretations of probability. *Erkenntnis*, **69**, 335–49.

Russo, Federica and Williamson, Jon (2007). Interpreting probability in causal models for cancer. In *Causality and Probability in the Sciences* (eds. F. Russo and J. Williamson), Texts in Philosophy, pp. 217–41. College Publications, London.

Salmon, Wesley C. (1967). *Foundations of Scientific Inference*. University of Pittsburgh Press, Pittsburgh.

——(1990). Rationality and objectivity in science, or Tom Kuhn meets Tom Bayes. In *Scientific Theories* (ed. C. Wade Savage), pp. 175–204. University of Minnesota Press, Minneapolis, MN. Minnesota Studies in the Philosophy of Science 14.

——(2005). *Reality and rationality*. Oxford University Press, Oxford. Edited by Phil Dowe and Merrilee H. Salmon.

Savage, L. J. (1954). *The foundations of statistics*. Wiley. New York. John Wiley & Sons.

Scott, Dana and Kraus, Peter (1966). Assigning probabilities to logical formulas. In *Aspects of Inductive Logic* (eds. J. Hintikka and P. Suppes), pp. 219–64. North-Holland, Amsterdam.

Seidenfeld, Teddy (1979). Why I am not an objective Bayesian. *Theory and Decision*, **11**, 413–40.

——(1986). Entropy and uncertainty. *Philosophy of Science*, **53**(4), 467–91.

——Schervish, M. J., and Kadane, J. B. (1990). When fair betting odds are not degrees of belief. *PSA: Proceedings of the Biennial Meeting of the Philosophy of Science Association*, **1990**, 517–24.

Shafer, Glenn (1996). *The art of causal conjecture*. MIT Press, Cambridge, MA.

——and Pearl, Judea (eds.) (1990). *Readings in uncertain reasoning*. Morgan Kaufmann, San Mateo, CA.

Shannon, Claude (1948). A mathematical theory of communication. *The Bell System Technical Journal*, **27**, 379–423 and 623–56.

Shimony, Abner (1973). Comment on the interpretation of inductive probabilities. *Journal of Statistical Physics*, **9**(2), 187–91.

—— (1985). The status of the principle of maximum entropy. *Synthese*, **63**, 35–53.

Skyrms, Brian (1985). Maximum entropy inference as a special case of conditionalization. *Synthese*, **63**, 55–74.

—— (1987). Coherence. In *Scientific Inquiry in Philosophical Perspective* (ed. N. Rescher), pp. 225–42. University Press of America, Lanham, MD.

Smith, Cedric A. B. (1961). Consistency in statistical inference and decision. *Journal of the Royal Statistical Society Series B*, **23**, 1–37. With discussion.

Sturgeon, Scott (2008). Reason and the grain of belief. *Noûs*, **42**(1), 139–65.

Teller, Paul (1973). Conditionalisation and observation. *Synthese*, **26**, 218–58.

Topsøe, F. (1979). Information theoretical optimization techniques. *Kybernetika*, **15**, 1–27.

Tribus, Myron (1961). *Thermostatics and thermodynamics: an introduction to energy, information and states of matter, with engineering applications*. Van Nostrand, Princeton, NJ.

—— (1969). *Rational descriptions, decisions and designs*. Pergamon Press, New York.

Uffink, Jos (1996). The constraint rule of the maximum entropy principle. *Studies in History and Philosophy of Modern Physics*, **27**, 47–79.

van Fraassen, Bas C. (1981). A problem for relative information minimizers in probability kinematics. *British Journal for the Philosophy of Science*, **32**, 375–79.

—— (1984). Belief and the will. *Journal of Philosophy*, **81**(5), 235–56.

—— (1987). Symmetries of personal probability kinematics. In *Scientific inquiry in philosophical perspective* (ed. N. Rescher), pp. 183–223. University Press of America, Lanham, MD.

—— (1989). *Laws and symmetry*. Clarendon Press, Oxford.

—— (2002). *The empirical stance*. Yale University Press, New Haven, CT.

—— (2009). On stance and rationality. *Synthese*, doi 10.1007/s11229-009-9520-1.

—— Hughes, R. I. G., and Harman, G. (1986). A problem for relative information minimisers, continued. *British Journal for the Philosophy of Science*, **37**, 453–75.

Venn, John (1866). *Logic of chance: an essay on the foundations and province of the theory of probability* (Second (1876) edn). Macmillan, London.

von Kries, Johannes (1886). *Die Principien der Wahrscheinlichkeits-Rechnung: Eine logische Untersuchung*. Mohr, Tübingen.

von Mises, Richard (1928). *Probability, statistics and truth* (Second (1957) edn). Allen and Unwin, London.

Walley, Peter (1991). *Statistical reasoning with imprecise probabilities*. Chapman and Hall, London.

Wheeler, Gregory and Williamson, Jon (2010). Evidential probability and objective Bayesian epistemology. In *Handbook of the Philosophy of Statistics* (eds. P. S. Bandyopadhyay and M. Forster). Elsevier, Amsterdam.

Williams, Peter M. (1980). Bayesian conditionalisation and the principle of minimum information. *British Journal for the Philosophy of Science*, **31**, 131–44.

Williamson, Jon (1999). Countable additivity and subjective probability. *British Journal for the Philosophy of Science*, **50(3)**, 401–16.

—— (2001). Foundations for Bayesian networks. In *Foundations of Bayesianism* (eds. D. Corfield and J. Williamson), pp. 75–115. Kluwer, Dordrecht.

—— (2002). Maximising entropy efficiently. *Electronic Transactions in Artificial Intelligence Journal*, **6**. www.etaij.org.

—— (2005a). *Bayesian nets and causality: philosophical and computational foundations*. Oxford University Press, Oxford.

—— (2005b). Objective Bayesian nets. In *We Will Show Them! Essays in Honour of Dov Gabbay* (eds. S. Artemov, H. Barringer, A. S. d'Avila Garcez, L. C. Lamb, and J. Woods), Volume 2, pp. 713–30. College Publications, London.

—— (2007a). Inductive influence. *British Journal for the Philosophy of Science*, **58**(4), 689–708.

—— (2007b). Motivating objective Bayesianism: from empirical constraints to objective probabilities. In *Probability and Inference: Essays in Honour of Henry E. Kyburg Jr.* (eds. W. L. Harper and G. R. Wheeler), pp. 151–79. College Publications, London.

—— (2008a). Objective Bayesian probabilistic logic. *Journal of Algorithms in Cognition, Informatics and Logic*, **63**, 167–83.

—— (2008b). Objective Bayesianism with predicate languages. *Synthese*, **163**(3), 341–56.

—— (2009a). Aggregating judgements by merging evidence. *Journal of Logic and Computation*, **19**, 461–73.

—— (2009b). Objective Bayesianism, Bayesian conditionalisation and voluntarism. *Synthese*, doi 10.1007/s11229-009-9515-y.

—— (2009c). Philosophies of probability. In *Handbook of the Philosophy of Mathematics of the Handbook of the Philosophy of Science volume 4* (ed. A. Irvine), pp. 493–533. North-Holland, Amsterdam.

Williamson, Timothy (2000). *Knowledge and its limits*. Oxford University Press, Oxford.

Index

X-divergence, 155
n-divergence, 91
n-entropy, 91

Actual frequencies, 23
Affine, 75
AGM postulates, 138
Ancestral, 47
Argument by derivation, 31
Argument by interpretation, 31
Arguments, 125
Atomic n-state, 90
Atomic state, 27

Bayes, Thomas, 10, 15, 24
Bayesian, 14
Bayesian conditionalisation, 3, 15, 53, 59
Bayesian confirmation theory, 15
Bayesian epistemology, 15
Bayesian interpretation of probability, 15
Bayesian net, 56, 108
Bayesian statistics, 15
Belief contraction, 138
Belief function, 27
Belief merging, 136, 138
Belief revision, 136, 137
Belief update, 138
Bernoulli, Jakob, 10, 12, 24
Bertrand's paradox, 152
Betting interpretation of rational belief, 32
Betting quotient, 32
Blind Minimum Entropy Principle, 56
Borel function, 154

C1, 39
C2, 39
C3, 43
C4, 45
C5, 45
C6, 47
Calculi, 10
Calibration, v, 15, 28, 42, 47, 91, 151
Carnap, Rudolf, 21
Cautious Monotony, 133
Chance constraints, 47
Closer, 91, 95, 103
Collective Rationality, 137

Commensurable, 102
Compatible, 47
Compilation, 112
Conclusion Conjunction, 133
Conditional bet, 87
Conditional probability, 27
Conditionalization, 2
Confirmation theory, 169
Conflicts, 125
Conservative, 76
Consistency-based merging operators, 145
Constraint graph, 113
Countable Additivity, 11, 21–23, 148–150
Counterfactual frequencies, 23
Cox's theorem, 33
Cox, Richard, 33
Credal net, 110
Cross entropy, 28
Cross entropy updating, 53, 59

Dag, 109
Data, 4, 140
De Finetti, Bruno, 2, 15
Decomposable, 132
Degree of support, 124, 125
Dempster–Shafer theory, 34
Density, 154
Derivation, argument by, 31
Dictatorships, 137
Dimension reduction, 108
Directed acyclic graph, 109
Discursive dilemma, 137
Doctrinal paradox, 137
Dutch book, 18, 33–35

Elementary outcome, 27
Empirically based subjective Bayesian
    epistemology, 15
EP-calibrated objective Bayesianism, 167, 168
Epistemic background, 4, 140
Epistemic metaphysics, 170, 171
Epistemic objectivity, 158
Epistemic rationality, 9
Epistemology, 10
Equidistant, 103
Equivocation, v, 16, 28, 49, 91, 151

Event space, 11
Evidence, 4
Evidence, most pertinent, 39, 45
Evidential certainties, 126
Evidential probability, 68, 126, 141, 166, 167, 169
Exchangeability, 19, 41, 82
Explication, 33
External, 121

Feigenbaum's bottleneck, 56
Field, 11
Finite Additivity, 18
Fit-for-purpose, 8
Foundational, 76
Full objectivity, 159

Gaifman's condition, 91
Greater entropy, 91

Harman, Gilbert, 9
Harper identity, 138
Howson, Colin, 84, 157

Iid, 40
Imprecise probability, 34
Improper prior, 152
Independence, 137
Inference process, 52
Infinitary Bayesian net, 115
Infinitary credal net, 115
Influence relation, 47
Input, 112, 113, 134
Integrity constraints, 138
Internal, 121
Interpretation, argument by, 31
Irrelevance, 47
Irrelevant, 47

Jaynes, Edwin, 2, 3, 24, 51, 152
Jeffreys, Harold, 21
Johnson, William Ernest, 20
Judy Benjamin problem, 77

Kelly gambling, 64
Keynes, John Maynard, 20, 155
Kolmogorov, Andrey, 40
Kullback–Leibler divergence, 28
Kyburg, Henry, 68, 127, 169

L1, 64
L2, 64
L3, 64
L4, 65
Language, 7

Laplace, Pierre Simon, marquis de, 19, 20, 24, 155
Law of excluded gambling systems, 41
Left Classical Equivalence, 133
Levi identity, 138
Lewis, David, 40, 85, 161
Logical interpretation of probability, 20
Logical objectivity, 11
Logical omniscience, 34, 38

Majority, 145
Markov Condition, 109, 110
Maxent, 25
Maximally equivocal evidentially compatible probability functions, 29
Maximum Entropy Principle, 3, 25, 26, 29, 33, 51–54, 56, 59–62, 67, 68, 71, 83, 92, 155–157
Miller's Principle, 39
Minimum Entropy Principle, 54, 56, 60
Mixed, 121
Monotonic, 132
Most pertinent evidence, 45

Non-conflicts, 125

Objective Bayesian epistemology, 1, 16
Objective Bayesian interpretation of probability, 10
Objective Bayesian net, 3, 108
Objective Bayesianism, 1
Objective credal net, 108, 115
Objectivity, 10
Obnet, 112
Ocnet, 115
Ontological objectivity, 11
Outcome space, 11
Output, 112, 114, 134

P1, 27, 35
P2, 27, 35
P3, 27, 35
Parameterized credal net, 111
Paris, Jeff, 51
Parsimony, 10
Partition, 41
Pearl, Judea, 4, 56
Pertinent, 39, 45
Posterior, 53
Potential equivocators, 152
PP1, 90, 93
PP2, 90, 93
PP3, 90, 93
PP4, 91, 93
Practical certainty, 126
Practical rationality, 9

Pragmatic encroachment, 9
Pragmatic rationality, 9
Precautionary Principle, 66
Preferential consequence relation, 132
Preferential model, 132
Premiss Disjunction, 133
Principal Principle, 39, 40, 161, 162, 169
Principle of Indifference, 4, 20, 23, 155, 157
Principle of Insufficient Reason, 20
Principle of Reflection, 87, 88
Principle of the Narrowest Reference Class, 13
Prior, 2, 53
Probabilistic variables, 125
Probability, v, 15, 27, 90, 149
Probability function, 11, 27, 90
Probability space, 11, 149
Progic, 121
Proportional gambling, 64

Query-answering, 112

Radical subjectivism, 73
Radical voluntarism, 73
Radon–Nikodym derivatives, 154
Radon–Nikodym theorem, 154
Ramsey, Frank, 15, 16, 32, 57
Ramsey-de Finetti theorem, 35
Random variable, 149
Rational consequence relations, 141
Reference class, 13
Reference prior, 152
Regular, 131
Relevance set, 47
Reliability threshold, 145
Respect, 154
Richness, 126
Right Weakening, 133
Robust Bayes, 63
Rosenkrantz, Roger, 25

Salmon, Wesley, 25
Sample frequencies, 23

Scoring rule, 64
Second-order evidential probability, 127, 167, 168
Sensitivity analysis, 66
Shannon, Claude, 51
Simple, 78
Simple random variable, 149
Single-case, 148
Smooth, 132
Specificity, 126
Stably equidistant, 103
Strength, 126
Strictly subjective Bayesian epistemology, 15
Strong law of large numbers, 40
Strong limit point, 96
Structural constraints, 47
Subjective Bayesian epistemology, 16
Sufficiently close, 30
Sufficiently equivocal evidentially compatible probability functions, 30
System-P, 132

Teller, Paul, 85
Theoretical rationality, 9
Total evidence, 4
Trigger function, 61

Ultimate belief, 24, 148, 159, 165
Unanimity, 137
Undermining, 161
Universal Domain, 137
Unstably equidistant, 104

Variety, 10
Vencovská, Alena, 51
Venn, John, 23

Weak law of large numbers, 13
Weak limit point, 96
Williamson, Timothy, 5
Wine-water paradox, 152